U.S. Marine Corps Close Combat

Plus

Combat Breaching

Department of Defense

Pentagon
Publishing

DEPARTMENT OF THE NAVY
Headquarters United States Marine Corps
Washington, D.C. 20380-1775

18 February 1999

FOREWORD

1. PURPOSE

Today's Marines operate within a continuum of force where conflict may change from low intensity to high intensity over a matter of hours. Marines are also engaged in many military operations other than war, such as peacekeeping missions or noncombatant evacuation operations, where deadly force may not be authorized. During noncombative engagements, Marines must determine if a situation warrants applying deadly force. Sometimes Marines must decide in a matter of seconds because their lives or the lives of others depend on their actions. To make the right decision, Marines must understand both the lethal and nonlethal close combat techniques needed to handle the situation responsibly without escalating the violence unnecessarily. Marine Corps Reference Publication (MCRP) 3-02B, *Close Combat,* provides the tactics, techniques, and procedures of Marine Corps close combat. It also provides the doctrinal basis for the Marine Corps Close Combat Training Program (MCCCTP).

2. SCOPE

This publication guides individual Marines, unit leaders, and close combat instructors in the proper tactics, techniques, and procedures for close combat. MCRP 3-02B is not intended to replace supervision by appropriate unit leaders and close combat instruction by qualified instructors. Its role is to ensure standardization and execution of tactics, techniques, and procedures throughout the Marine Corps.

3. SUPERSESSION

MCRP 3-02B supersedes Fleet Marine Force Manual (FMFM) 0-7, *Close Combat,* dated 9 July 1993. There are significant differences between the two publications. MCRP 3-02B should be reviewed in its entirety.

4. WARNING

Techniques described in this publication can cause serious injury or death. Practical application in the training of these techniques will be conducted in strict accordance with approved Entry Level Close Combat, Close Combat Instructor (CCI), and Close Combat Instructor Trainer (CCIT) lesson plans. Where serious danger exists, the reac er is alerted by the following:

WARNING

5. CERTIFICATION

Reviewed and approved this date.

BY DIRECTION OF THE COMMANDANT OF THE MARINE CORPS

J. E. RHODES
Lieutenant General, U.S. Marine Corps
Commanding General
Marine Corps Combat Development Command

DISTRIBUTION: 14400006600

OVERVIEW OF CLOSE COMBAT

1. Purpose of Close Combat

Close combat is the physical confrontation between two or more opponents. It involves armed and unarmed and lethal and nonlethal fighting techniques that range from enforced compliance to deadly force. The purpose of close combat is to execute armed and unarmed techniques to produce both lethal and nonlethal results. Unarmed techniques include hand-to-hand combat and defense against hand-held weapons. Armed techniques include techniques applied with a rifle, bayonet, knife, baton, or any weapon of opportunity.

2. Continuum of Force

Marines will find themselves in both combative and noncombative situations. The threat level in these situations can rise and fall several times based on the actions of both Marines and the people involved. The escalation of force stops when one person complies with the demands imposed by another person. This range of actions is known as a continuum of force. Continuum of force is the concept that there is a wide range of possible actions, ranging from voice commands to application of deadly force, that may be used to gain and maintain control of a potentially dangerous situation (MCO 5500.6_, *Arming of Security and Law Enforcement [EE] Personnel and the Use of Force).* Continuum of force consists of five levels that correspond to the behavior of the people involved and the actions Marines use to handle the situation (see the table below). Close combat techniques are executed in levels three, four, and five.

Level One: Compliant(Cooperative)

The subject complies with verbal commands. Close combat techniques do not apply.

Level Two: Resistant (Passive)

The subject resists verbal commands but complies immediately to any contact controls. Close combat techniques do not apply.

Continuum of Force		
Level	**Description**	**Actions**
1	Compliant (Cooperative)	Verbal commands
2	Resistant (Passive)	Contact controls
3	Resistant (Active)	Compliance techniques
4	Assaultive (Bodily Harm)	Defensive tactics
5	Assaultive (Serious Bodily Harm/ Death)	Deadly force
Note: Shading indicates levels in which Marines use close combat techniques.		

Level Three: Resistant (Active)

The subject initially demonstrates physical resistance. Marines use compliance techniques to control the situation. Level three incorporates close combat techniques to physically force a subject to comply. Techniques include—

- Come-along holds.
- Soft-handed stunning blows.
- Pain compliance through joint manipulation and the use of pressure points.

Level Four: Assaultive (Bodily Harm)

The subject may physically attack Marines, but he does not use a weapon. Marines use defensive tactics to neutralize the threat. Defensive tactics include the following close combat techniques:

- Blocks.
- Strikes.
- Kicks.
- Enhanced pain compliance procedures.
- Nightstick blocks and blows.

Level Five: Assaultive (Serious Bodily Harm/Death)

The subject usually has a weapon and will either kill or seriously injure someone if he is not stopped immediately and brought under control. Typically, to control the subject, Marines apply deadly force through the use of a firearm, but they may also use armed and unarmed close combat techniques.

3. Marine Corps Tactical Concepts

Close combat techniques support the following key Marine Corps tactical concepts. The concepts are not standalone ideas but are to be combined to achieve an effect that is greater than their separate sum.

Achieving a Decision

Achieving a decision is important in close combat. An indecisive fight wastes energy and possibly Marines' lives. Whether the intent is to control an opponent through restraint or defend themselves in war, Marines must have a clear purpose before engaging in close combat and act decisively once engaged.

Gaining an Advantage

A basic principle of martial arts is to use the opponent's strength and momentum against him to gain more leverage than one's own muscles alone can generate, thereby gaining an advantage. In close combat, Marines must exploit every advantage over an opponent to ensure a successful outcome. This can include employing various weapons and close combat techniques that will present a dilemma to an opponent. Achieving surprise can also greatly increase leverage. Marines try to achieve surprise through deception, stealth, and ambiguity.

Speed

Marines use speed to gain the initiative and advantage over the enemy. In close combat, the speed and violence of the attack against an opponent provides Marines with a distinct advantage. Marines must know and understand the basics of close combat so they can act instinctively with speed to execute close combat techniques.

Adapting

Close combat can be characterized by friction, uncertainty, disorder, and rapid change. Each situation is a unique combination of shifting factors that cannot be controlled with precision or certainty. For example, a crowd control mission may call for Marines to employ various techniques ranging from nonlethal restraint to more forceful applications. Marines who adapt quickly will have a significant advantage.

Exploiting Success

Typically, an enemy will not normally surrender simply because he was placed at a disadvantage. Marines cannot be satisfied with gaining an advantage in a close combat situation. They must exploit any advantage aggressively and ruthlessly until an opportunity arises to completely dominate the opponent. Marines must exploit success by using every advantage that can be gained.

(reverse blank)

Close Combat

Table of Contents

Page

Chapter 5. Throws

Chapter 6. Chokes and Holds

Chapter 7. Ground Fighting

Chapter 8. Nonlethal Techniques

Appendices

CHAPTER 1

FUNDAMENTALS OF CLOSE COMBAT

This chapter describes all techniques for a right-handed person. However, all techniques can be executed from eittwr side.

The Marine is depicted in camouflage utilities. The opponent is depicted without camouflage.

The fundamentals of close combat include ranges, weapons of the body, target areas of the body, and pressure points of the body. These fundamentals form the basis for all close combat techniques. They provide Marines with a common framework regardless of the type of confrontation or the techniques used. If Marines apply these fundamentals properly in a close combat situation, they may save their lives or the lives of fellow Marines.

1. Ranges of Close Combat

Close combat engagements occur within three ranges: long range, midrange, and close range. During any engagement, these ranges may blur together or may rapidly transition from one to another until either the opponent is defeated or the conflict is resolved.

Long Range

During long range engagements, combatants engage each other with rifles, bayonets, sticks, or entrenching tools. See figure below.

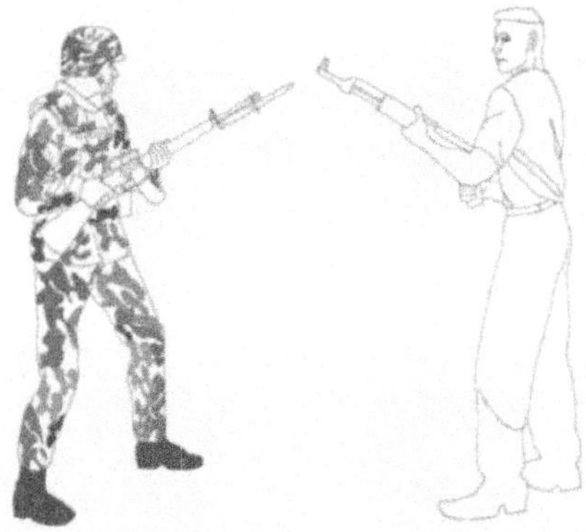

Midrange

During midrange engagements, combatants engage each other with knives, punches, or kicks.

Close Range

During close range engagements, combatants grab each other. Close range engagements also involve elbow strikes, knee strikes, and grappling.

2. Weapons of the Body

Hands and Arms

The hands, forearms, and elbows are the arm's individual weapons. The hands consist of several areas that can be used as weapons: fists, edges of hands, palms, and fingers.

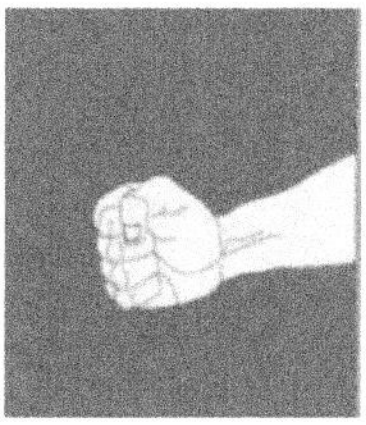

Fists. To minimize injury to the fists, Marines use their fists as weapons to target soft tissue areas such as the throat. The fists' striking surfaces are the first two knuckles of the hands or the meaty portions of the hands below the little fingers.

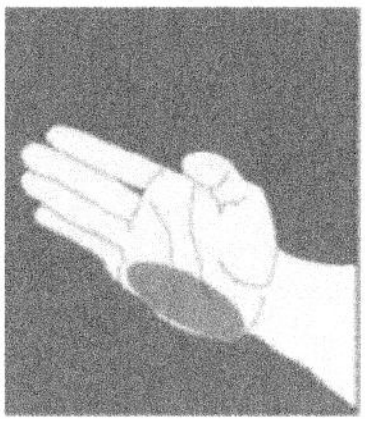

Edge of Hand. Marines use the edge of the hand (knife edge) as a weapon. Marines use the edge of the hand to strike soft tissue areas.

Palms. Because of the palm's padding, Marines use the heels of the palms to strike, parry, and/or block.

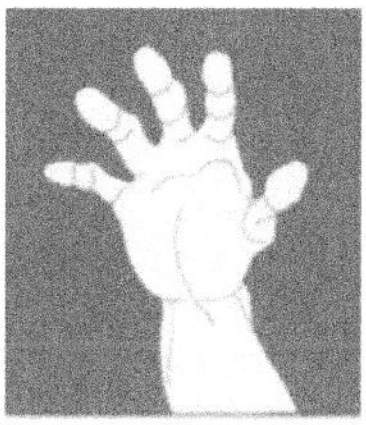

Fingers. Marines use the fingers to gouge, rip, and tear soft tissue areas (e.g., eyes, throat, groin).

Forearms. Marines use the forearms as a defensive tool to deflect or block attacks. Forearms can also be used as striking weapons to damage or break an opponent's joints and limbs. Marines sustain less self-injury when strikes are conducted with the forearms than when strikes are conducted with fists and fingers.

Elbows. Marines use the elbows as striking weapons. Because of the short distance needed to generate power, elbows are excellent weapons for striking during the close range of close combat.

Legs

The legs are more powerful than any other weapon of the body, and they are less prone to injury when striking. The feet are protected by boots and are the preferred choice for striking.

Feet. Marines use the balls of the feet, the insteps, and the toes to kick an opponent. Marines use the cutting edge of the heels and the heels to stomp on an opponent. Marines must be wearing boots when striking with the toes.

Knees. Like elbows, knees are excellent weapons in the close range of close combat. Knee strikes are most effective while fighting close to an opponent where kicks are impractical. The opponent's groin area is an ideal target for the knee strike if he is standing upright. Knee strikes can deliver a devastating secondary attack to an opponent's face following an initial attack that caused him to bend at the waist.

3. Target Areas of the Body

During close combat, Marines strive to attack the accessible target areas of an opponent's body. The readily accessible areas will vary with each situation and throughout the engagement. The target areas are divided into five major groups: head, neck, torso, groin, and extremities. The figure below illustrates target areas of the body.

Head

The vulnerable regions of the head are the eyes, temple, nose, ears, and jaw. Massive damage to the head kills an opponent.

Eyes. The eyes are excellent targets because they are soft tissue areas that are not protected by bone or muscle. Attacks to this area may cause the opponent to protect the area with his hands, allowing Marines to execute a secondary attack to other

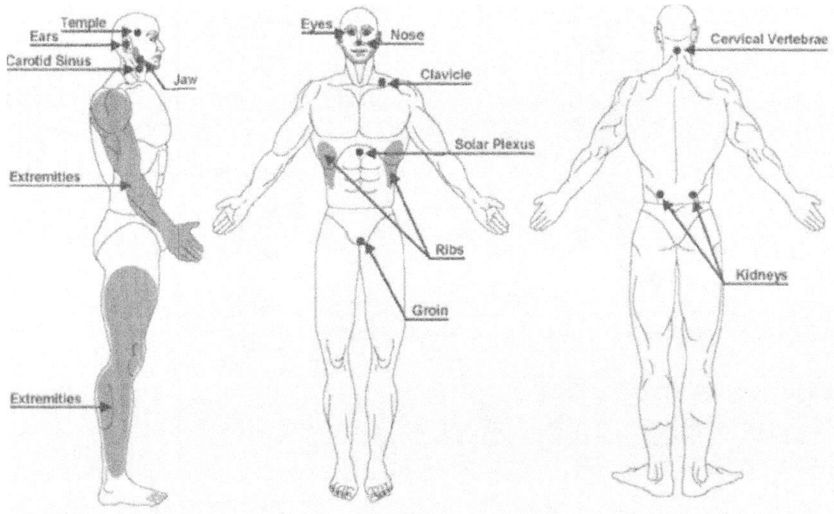

target areas while the opponent uses his hands to protect his eyes.

Temple. The temple is one of the most fragile areas of the head. Powerful strikes to the opponent's temple cause permanent damage and death.

Nose. The nose is very sensitive and easily broken. An attack to this area causes involuntary watering and closing of the opponent's eyes, rendering him vulnerable to secondary attacks. However, through training, individuals can condition themselves to withstand attacks to the nose. Therefore, any attack to the nose must be powerfully delivered.

Ears. Attacks to the ears may cause the eardrum to rupture. But this may not stop or even distract an opponent unless Marines powerfully deliver the strike.

Jaw. The jaw region, when struck forcefully, renders the opponent unconscious. Strikes to the jaw cause painful injuries to the teeth and surrounding tissues (e.g., lips, tongue), but the risk of self-injury is great unless Marines deliver strikes with a hard object such as a helmet, rifle butt, or boot heel.

Neck

The front of the neck, or throat area, is a soft tissue area that is not covered by natural protection. Damage to this region causes the opponent's trachea to swell, closing his airway, which can lead to death.

Carotid Sinus. The carotid sinus is located on both sides of the neck just below the jaw. Strikes to the carotid sinus restrict blood flow to the brain, causing loss of consciousness or death.

Cervical Vertebrae. The cervical vertebrae on the back of the neck, from the base of the skull to the top of the shoulders, contains the spinal cord, which is the nervous system's link to the brain. The weight of the head and the lack of large muscle mass allow damage to the cervical vertebrae and spinal cord. Excessive damage to this area causes pain, paralysis, or death.

Torso

Clavicle. The opponent's clavicle (or collar bone) can be easily fractured, causing immobilization of the arm.

Solar Plexus. Attacks to the opponent's solar plexus or center of the chest can knock the breath out of him and immobilize him.

Ribs. Damage to the opponent's ribs immobilizes him. It may also cause internal trauma.

Kidneys. Powerful attacks to the opponent's kidneys cause immobilization, permanent damage, or death.

Groin

The groin area is another soft tissue area not covered by natural protection. Any damage to this area causes the opponent to involuntarily protect his injured area, usually with his hands or legs. In male opponents, the scrotum is the main target since even a near miss causes severe pain, contraction of the lower abdominal muscles, deterioration of his stance, and possible internal trauma.

Extremities

Rarely will an attack to the opponent's extremities (arms and legs) cause death, but they are important target areas in close combat. Damage to an opponent's joints causes immobilization.

4. Pressure Points of the Body

There are nerves in the human body that, when pressure is applied or when they are struck, allow Marines to control a subject through pain compliance. Marines use pressure points to control an opponent when deadly force is not authorized. They also use pressure points to soften or distract an opponent so a lethal or nonlethal technique can be employed. The figure on page 1-5 illustrates the body's pressure points. Marines execute attacks to pressure points by—

: Rapidly kicking or striking pressure points.
: Slowly applying steady pressure to pressure points.

Infraorbital Nerve

The infraorbital nerve is just below the nose. Marines apply pressure to this nerve with an index finger to control the opponent.

Mastoid Process

The mastoid process is behind the base of the ear and beneath the edge of the jaw. Marines apply inward and upward pressure to this pressure point with the fingers to distract and control the opponent.

Jugular Notch

The jugular notch is at the base of the neck in the notch formed at the center of the clavicle. Marines apply pressure in a quick, stabbing motion with the index finger. Strikes to the jugular notch cause serious damage.

Brachial Plexus (Tie In)

The brachial plexus (tie in) is on the front of the shoulder at the joint. Strikes and pressure applied with the hand are effective on this nerve.

Radial Nerves

Radial nerves are on the inside of the forearms along the radius bones. Strikes and pressure applied with the hand to the radial nerve serve as a softening technique.

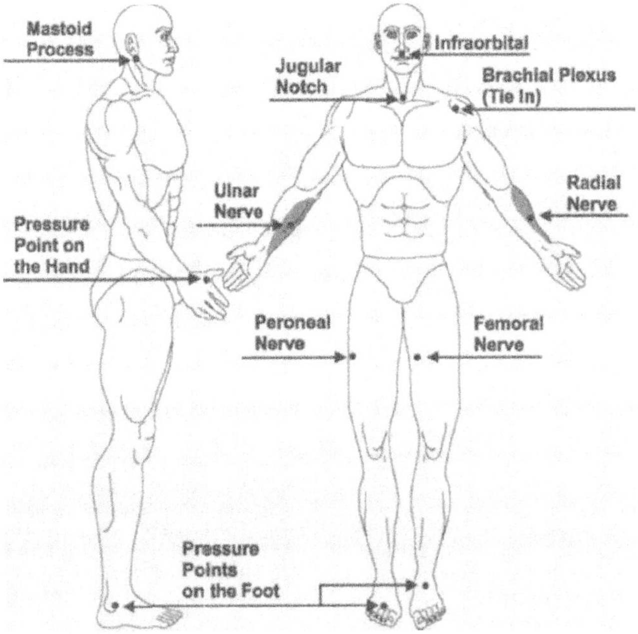

Ulnar Nerve

Ulnar nerves are on the outside of the forearms along the ulnar bones. Strikes and pressure applied with the hand to the ulnar nerve serve as a softening technique.

Pressure Point on the Hand

The hands contain a pressure point on the webbing between the thumbs and index fingers where the two bones of the fingers meet.

To force an opponent to soften or release his grip, Marines apply pressure with their index fingers to this pressure point or strike this pressure point with their fists.

Femoral Nerves

Femoral nerves are on the inside of the thighs along the femur bones. Strikes to the femoral nerve serve as a softening technique.

Peroneal Nerves

Peroneal nerves are on the outside of the thighs along the femur bones. Strikes to the peroneal nerve serve as a softening technique.

Pressure Points on the Feet

There are pressure points on the feet that, when pressure is applied or when they are struck, serve to soften or distract the opponent. Marines apply pressure with the toe, edge, or heel of their boots to the following points:

: The notch below the ball of the ankle.
: The top center of the foot, above the toes.
: The top of the foot where the leg and foot meet.

5. Basic Warrior Stance

Achieving a solid stance is critical to stability and movement throughout any close combat confrontation. The basic warrior stance provides the foundation for stability and movement that is needed to execute close combat techniques. To execute the basic warrior stance, Marines put their feet apart, hands up, elbows in, and chin down.

Feet Apart

Place feet shoulder-width apart.

Keep the head forward and eyes on the opponent, take a half step forward with the left foot, and pivot on the heels so the hips and shoulders are at approximately a 45-degree angle to the right.

Distribute body weight evenly on both legs. Bend the knees slightly.

Hands Up

Curl the fingers naturally into the palm of the hand. Position the thumb across the index and middle fingers. Do not clench the fists. Clenching the fists increases muscular tension in the forearms and decreases speed and reaction time.

Bring the hands up to the face at chin level, with the palms facing each other. Hold the fists up high enough to protect the head, but not so high that they block the field of vision. Ensure continuous eye contact with the opponent.

Elbows In

Tuck the elbows in close to protect the body.

Chin Down

Tuck the chin down to take advantage of the natural protection provided by the shoulders.

6. Angles of Approach and Movement

Marines use movement to control a confrontation and to retain a tactical advantage. Movement increases power and maximizes momentum. By moving around the opponent, Marines gain access to different target areas of the opponent's body. Movement allows Marines to use different weapons of the body and different close combat techniques to attack specific target areas.

Angles of Approach

Marines move anywhere within a 360-degree circle around the opponent to gain a tactical advantage. This circle provides access to different target areas of the opponent's body.

When facing an opponent, Marines move in a 45-degree angle to either side of the opponent. Moving at a 45-degree angle avoids an opponent's strike and puts Marines in the best position to attack the opponent. Marines should avoid being directly in front of an opponent during a confrontation. If a Marine is directly in front of an opponent, the opponent can rely on his forward momentum and linear power to seize the tactical advantage.

Movement

Marines must know how to move in all directions while maintaining the basic warrior stance. During any movement, the legs or feet should not be crossed. Once a movement is completed, the basic warrior stance should be resumed. Maintaining the basic warrior stance protects Marines and puts them in the proper position to launch an attack against an opponent.

Note: Before body movement begins, Marines turn their heads quickly to the new direction. The faster the head turns, the faster the body moves, and the quicker Marines attain visual contact with the opponent.

Forward to the Left. To move forward to the left from the basic warrior stance, Marines-

: Move the left foot forward at a 45-degree angle from the body (approximately 12 to 15 inches), keeping the toe pointed toward the opponent.

Bring the right foot behind the left foot as soon as the left foot is in place. This returns Marines to the basic warrior stance.

Forward to the Right. To move forward to the right from the basic warrior stance, Marines-

Move the right foot forward at a 45-degree angle from the body (approximately 12 to 15 inches).

Bring the left foot, toe pointing toward the opponent, in front of the right foot as soon as the right foot is in place. This returns Marines to the basic warrior stance.

Backward to the Left. To move backward to the left from the basic warrior stance, Marines execute the forward movement in reverse. Marines-

Move the left foot backward at a 45-degree angle from the body (approximately 12 to 15 inches), keeping the toe pointed toward the opponent.

Bring the right foot behind the left foot as soon as the left foot is in place. This returns Marines to the basic warrior stance.

Backward to the Right. To move backward to the right from the basic warrior stance, Marines execute the forward movement in reverse. Marines-

Move the right foot backward at a 45-degree angle from the body (approximately 12 to 15 inches).

Bring the left foot, toe pointing toward the opponent, in front of the right foot as soon as the right foot is in place. This returns Marines to the basic warrior stance.

7. Balance and Off-Balancing

Balance

In any close combat situation, Marines must strive to maintain balance. The last place to be in a close combat situation is on the ground. Marines must maintain a strong base and a low center of balance, their feet must be a shoulder-width apart, and they must stay on their toes to enable quick movement.

Off-balancing

Marines use off-balancing techniques to control an opponent. These techniques are used to throw an opponent to the ground while Marines remain standing, or they are used to put Marines in a position for an offensive attack.

Off-balancing techniques use the opponent's momentum to move or throw him. For example, if the opponent is charging a Marine, the Marine pulls the opponent to drive him to the ground. Likewise, if the opponent is pulling a Marine, the Marine pushes the opponent to drive him to the ground.

Off-balancing techniques also rely on the power generated by the opponent. For example, during combat a Marine may be tired or outnumbered. Depending on the generated energy and momentum of the opponent, the Marine employs off-balancing techniques with very little effort and still provides effective results.

Because off-balancing techniques rely on the momentum and power generated by the opponent, these techniques are particularly effective for Marines who may be outsized by their opponent or lack their opponent's strength.

Angles of Off-balancing

There are eight angles or directions in which an opponent can be off-balanced: forward, rear, right, left, forward right, forward left, rear right, and rear left.

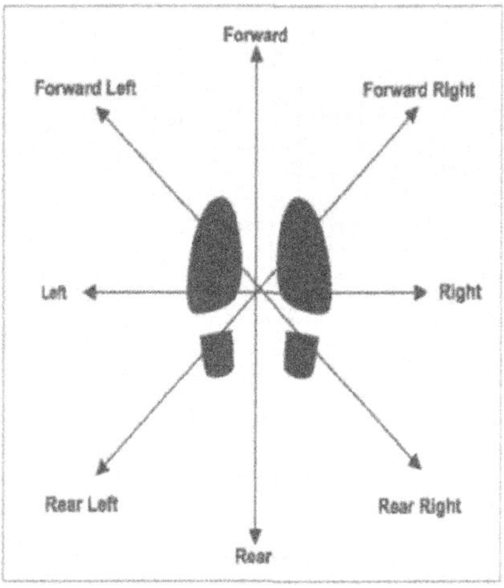

Note: The angles correspond to the Marine's perspective, not the opponent's. Forward, rear, right, and left are straight angles. Forward right, forward left, rear right, and rear left are considered quadrants that are at a 45-degree angle in either direction to the front or rear.

Off-balancing Techniques

Marines off-balance an opponent by pushing, pulling, or bumping the opponent with their hands, arms, or bodies.

To pull, Marines grab an opponent with their hands and drive him forcefully to one of the rear quadrants or to the right or left.

To push, Marines grab the opponent with their hands and drive him forcefully into one of the front quadrants or to the right or left.

Marines execute bumping in the same manner as pushing, but use their shoulders, hips, and legs instead of their hands to off-balance the opponent.

8. Falls

Marines may lose their balance or be thrown to the ground during encounters with an opponent. Marines use falling techniques to absorb the impact of a fall and to quickly return to their feet following an opponent's attack.

Whether falling or being thrown by an opponent, Marines strive to reduce the force of the impact, to prevent serious personal injury, and to increase the chances of survival. Falling techniques use the body's large muscles (back, thighs, buttocks) to protect vital organs and bones from injury and immobilization.

Front Fall

Marines execute a front fall to break a fall on the front. To execute the front fall, Marines-

Bend the elbows and place the palms facing out in a position to spread and absorb the impact of the fall.

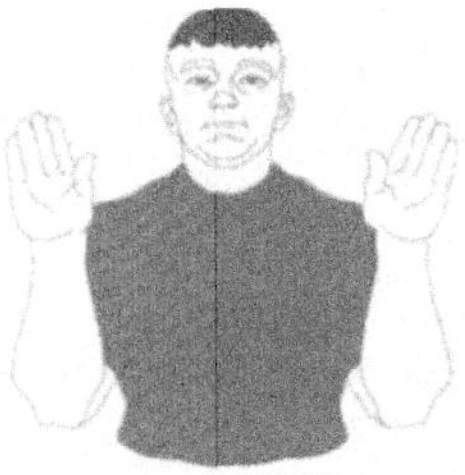

Fall forward, breaking the fall with the forearms and palms. The forearms and hands, down to the fingertips, should strike the ground simultaneously.

Offer resistance with the forearms and hands to keep the head raised off the ground.

Side Fall

Marines execute a side fall to break a fall on the side. To execute the side fall, Marines-

Bring the right arm across the body so the hand is next to the left shoulder with the palm facing inboard.

Fall to the side, breaking the fall with the right arm by slapping the ground and making contact from the shoulder or forearm down to the hand. At the same time, tuck the chin and keep the head raised off the ground. The chin should be tucked to the chest at all times to prevent whiplash.

Stretch the right leg out to make contact with the ground and to distribute and absorb the impact. Bend the left leg, allowing the foot to make contact with the ground.

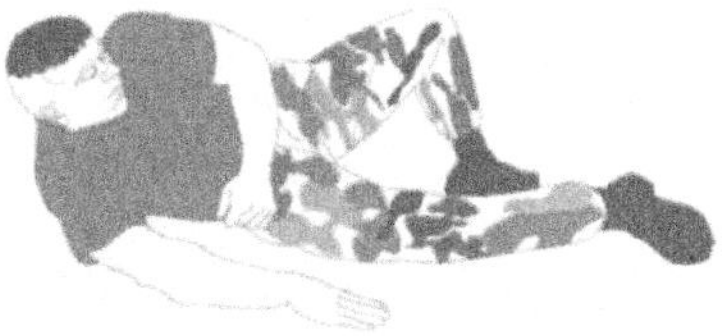

Back Fall

Marines execute a back fall to break the fall when being thrown or falling backward. To execute the backfall, Marines-

Cross the hands in front of the chest and tuck the chin.

Fall backward and slap the ground with the fore-arms and hands to absorb the impact of the fall and keep the head off the ground.

Forward Shoulder Roll

Marines use the forward shoulder roll to break a fall from an opponent's attack and to use the momentum of the fall to get back on their feet quickly. Ideally, Marines execute the forward shoulder roll to a standing position so they can continue fighting. To execute the forward shoulder roll, Marines-

Contact the ground with the back of the right fore-arm and upper arm. Tuck the chin into the chest.

Roll onto the right shoulder, rolling diagonally across the back to land on the left hip.

Slap the ground with the left arm, absorbing the impact from the shoulder to the hand, palm down.

Keep the left leg straight to absorb as much of the impact as possible. The right leg is bent and the foot hits flat on the ground.

Bend the left leg upon impact to push off with the left knee and leg to a squatting and then a standing position. Forward momentum should carry the Marine to a standing position.

CHAPTER 2

LETHAL AND NONLETHAL WEAPONS TECHNIQUES

This chapter describes all techniques for a right-handed person. However, all techniques can be executed from either side.

In drawings, the Marine is depicted in woodland camouflage utilities; the opponent is depicted without camouflage. In photographs, the Marine is depicted in woodland camouflage utilities; the opponent is depicted in desert camouflage utilities.

1. Bayonet Techniques

WARNING

During training, Marines must have bayonets sheathed. Marines use bayonet dummies to practice bayonet techniques. When practicing offensive and defensive bayonet techniques student-on-student, Marines use pugil sticks.

All Marines armed with a rifle carry a bayonet. The bayonet is an effective weapon if Marines are properly trained in offensive and defensive bayonet techniques. An offensive attack, such as a thrust, is a devastating attack that can quickly end a fight. Defensive techniques, such as the block and parry, can deter the opponent's attack and allow Marines to regain the initiative. Through proper training, Marines develop the courage and confidence required to effectively use a bayonet to protect themselves and destroy the enemy. In situations where friendly and enemy troops are closely mingled and rifle fire and grenades are impractical, the bayonet becomes the weapon of choice.

Holding the Rifle

To execute bayonet techniques, Marines hold the rifle in a modified basic warrior stance. All movement begins and ends with the basic warrior stance. To hold the rifle, Marines—

Use an overhanded grasp to grab the small of the rifle's stock. Use an underhanded grasp to grab the handguards of the rifle.

Lock the buttstock of the rifle against the hip with the right forearm.

Orient the blade end of the rifle toward the opponent.

Offensive Bayonet Techniques

Straight Thrust. Marines use the straight thrust to disable or kill an opponent. It is the most deadly offensive technique because it causes the most trauma to an opponent. Target areas are the opponent' s throat, groin, or face. The opponent's chest and stomach are also excellent target areas if not protected by body armor or combat equipment. To execute the straight thrust, Marines—

Lift the left leg and lunge forward off the ball of the right foot while thrusting the blade end of the weapon forward, directly toward the opponent.

Retract the weapon and return to the basic warrior stance.

Slash. Marines use the slash to kill an opponent or to create an opening in his defense. The target area is the opponent's neck. To execute the slash, Marines—

Extend the left hand back toward the left shoulder.

Thrust the left hand forward and swing it to the right, bring the right hand back toward the hip, and turn the cutting edge of the blade toward the opponent's neck. The movement is a slashing motion so the blade cuts across the opponent's neck.

Horizontal Buttstroke. Marines use the horizontal buttstroke to weaken an opponent's defenses, to cause serious injury, or to set him up for a killing blow. Target areas are the opponent's head, neck, and legs. To execute the horizontal buttstroke, Marines—

Step forward with the right foot and drive the right hand forward. Rotate the hips and shoulders into the strike. Move the left hand back toward the left shoulder.

Strike the opponent with the butt of the weapon.

Vertical Buttstroke. Marines use the vertical buttstroke to weaken an opponent's defenses, to cause serious injury, or to set him up for a killing blow. Target areas are the opponent's groin and face. To execute the vertical buttstroke, Marines—

Step forward with the right foot and drive the right hand straight up.

Pull the left hand back over the left shoulder.

Strike the opponent with the butt of the weapon.

Smash. Marines use the smash as a follow-up technique to the horizontal or vertical buttstroke, particularly if they missed the target. The target area is the opponent's head. To execute the smash following a buttstroke, Marines-

Step forward with the right foot and place the blade end of the weapon over the left shoulder and elevate the right elbow above the shoulders.

Drive the arms straight forward, striking the opponent with the butt of the weapon.

Defensive Bayonet Techniques

Parry. Marines use a parry as a defensive technique to redirect or deflect an attack. A parry is a slight redirection of a linear attack by an opponent; e.g., a straight thrust or a smash. To execute the parry, Marines-

Use the bayonet end of the rifle to redirect the barrel or bayonet of the opponent's weapon.

Lock the weapon against the hip with the right forearm.

Rotate to the right or left, moving the bayonet end of the rifle to parry the opponent's attack. Rotation is generated from the hips.

Redirect or guide the opponent's weapon away from the body by exerting pressure against the opponent' s weapon.

High Block. Marines execute a high block against a vertical attack coming from high to low. To execute the high block, Marines—

Thrust the arms up forcefully at approximately a 45-degree angle from the body. The weapon should be over the top of the head and parallel to the ground. The elbows are bent, but there is enough muscular tension in the arms to absorb the impact and deter the attack.

Low Block. Marines execute the low block

against a vertical attack coming from low to high. To execute the low block, Marines—

Thrust the arms down forcefully at approximately a 45-degree angle from the body. The weapon

should be at or below the waist and parallel to the ground. The elbows are bent, but there is enough muscular tension in the arms to absorb the impact and deter the attack.

Left and Right Block. Marines execute a left or right block against a horizontal buttstroke or a slash. To execute the left or right block, Marines-

Thrust the arms forcefully to the right or left, holding the rifle vertically in the direction of the attack. The elbows are bent, but there is enough muscular tension in the arms to absorb the impact and deter the attack.

Counter Action Following the Block. After deflecting an opponent's attack with a block, Marines counter with a slash or a horizontal buttstroke to regain the initiative. However, the objective in any bayonet fight is to thrust forward with the blade end of the weapon to immediately end the fight.

Group Strategy

On occasion, Marines may engage an opponent as a member of a group or numerous opponents by one's self or as a member of a group. By combining bayonet fighting movements and simple strategies, Marines can effectively overcome their opponent or opponents.

Offensive Strategy: Two Against One. If two bayonet fighters engage one opponent, the fighters advance together.

Fighter 1 engages the opponent while fighter 2 swiftly and aggressively attacks the opponent's exposed flank and destroys the opponent.

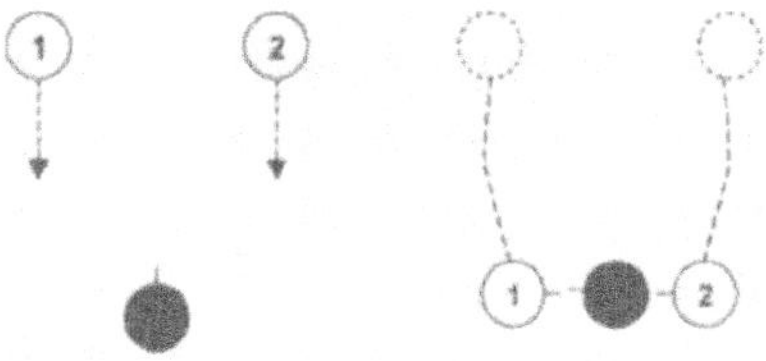

Offensive Strategy: Three Against Two. If three bayonet fighters engage two opponents, the fighters advance together keeping their opponents to the inside.

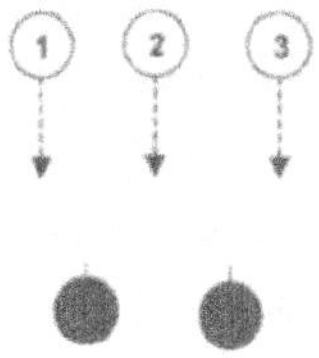

Fighters 1 and 3 engage opponents. Fighter 2 attacks the opponent's exposed flank engaged by fighter 1 and destroys the opponent.

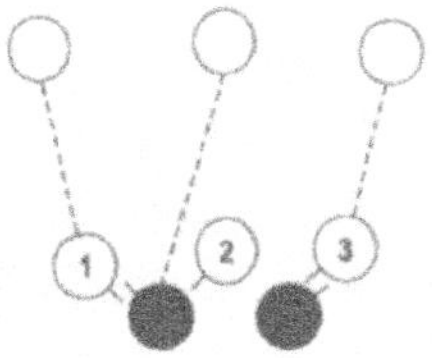

Fighters 1 and 2 turn and attack the exposed flank of the opponent engaged by fighter 3 and destroy the opponent.

Defensive Strategy: One Against Two. If a fighter is attacked by two opponents, the fighter immediately positions himself at the flank of the nearest opponent and keeps that opponent between himself and the other opponent.

Using the first opponent's body as a shield against the second opponent, the fighter destroys the first opponent quickly before the second opponent moves to assist.

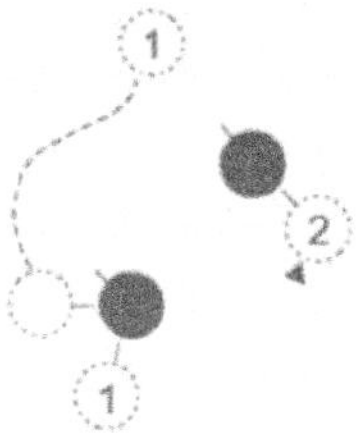

Then, the fighter engages and destroys the second opponent.

Defensive Strategy: Two Against Three. If two fighters are attacked by three opponents, the fighters immediately move to the opponent's flanks.

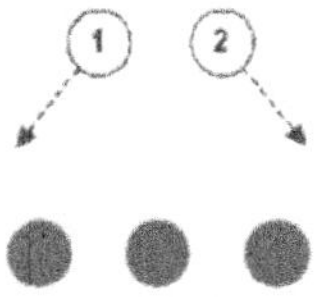

Fighters 1 and 2 quickly attack and destroy their opponents before the third opponent closes in.

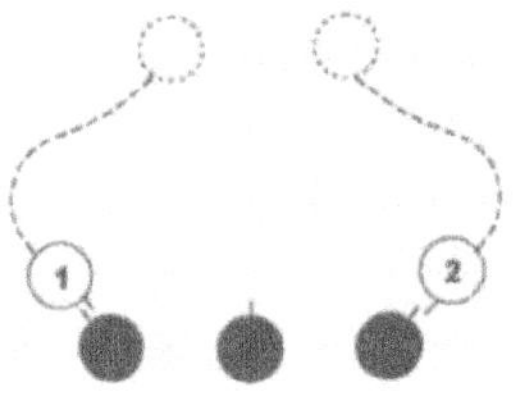

Fighter 1 engages the third opponent while fighter 2 attacks the opponent's exposed flank and destroys the opponent.

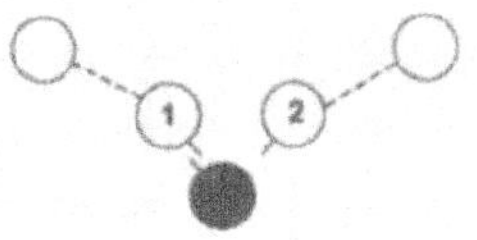

2. Nonlethal Rifle and Shotgun Retention Techniques

Most Marines are armed with the M16A2 service rifle. Marines are taught to keep their weapon with them at all times. Marines must be constantly alert to their surroundings and the people moving in and around their environment. Marines may be confronted by an individual who tries to take theii weapons. If this happens, Marines should not struggle with the individual. To retain positive control of their weapons, Marines must understand and apply weapons retention techniques, otherwise known as armed manipulation. The fol lowing techniques can be used with either the rifle or the shotgun.

Blocking Technique

To execute a blocking technique, Marines-

Stand in a defensive posture.

Use the weapon to block the opponent by thrusting it out firmly, with the elbows bent. Do not try to hit the opponent with the rifle. The rifle is used as a barrier.

Technique if Opponent Grabs Weapon Underhanded

Usually an opponent will try to grab the weapon or block it as an instinctive action. If the opponent uses an underhand grab to seize the handguards of the weapon, Marines—

Trap the opponent's closest finger(s) above the knuckle with the thumb so he cannot release his grip-

Apply bone pressure on the opponent's finger to initiate pain compliance.

Rotate the barrel of the weapon up or down quickly while maintaining pressure on the opponent's hand. At the same time, quickly pivot to off-balance the opponent.

Technique if Opponent Grabs Weapon Overhanded

If the opponent uses an overhand grab to seize the weapon overhanded, Marines—

Trap the opponent's finger to hold his hand in place.

Rotate the barrel to place it across the opponent's forearm and apply downward pressure. This ac-

tion is similar to an armbar.

Technique if the Opponent Grabs the Muzzle

If the opponent grabs the muzzle of the weapon, Marines—

Rotate the muzzle quickly in a circle motion, Slash downward with the muzzle to release the opponent's grip.

Butt Strikes

Strikes with the butt of the weapon control or ward off an attacker. During any of the retention techniques, Marines use the heel or cutting edge of the weapon to deliver butt strikes to the inside or the outside of the opponent's thigh. The inside butt strike targets the femoral nerve as illustrated in the photo to the left.

The outside butt strike targets the peroneal nerve as illustrated in the photo to the left. Strikes can be made to the outside or inside of the thighs. If a strike to one side of the thigh misses, Marines follow back through with the butt of the weapon on the other side of the thigh.

Off-balancing Techniques

Marines apply off-balancing techniques to throw an opponent to the ground and retain possession of the weapon.

If the opponent grabs the weapon and pushes, Marines should not push on the weapon. They should—

Move with the momentum and movement of the

opponent by pivoting in the direction of the movement by stepping back.

Throw the opponent to the ground with a quick jerking movement by lowering the mu/./le and swinging the butt of the weapon.

If the opponent grabs the weapon and pulls, Marines—

Step on the opponent's foot and push forward to off-balance him and drive him to the ground.

Sweep the opponent's feet out from under him by hooking his leg with the leg and kicking backward.

3. Nonlethal Handgun Retention Techniques

Many Marines are armed with the M9 service pistol. Marines must keep their weapons in their possession at all times. Marines must be constantly alert to their surroundings and the people moving in and around the environment. Marines may be confronted by an individual who tries to take their weapons. To retain positive control of the weapon, Marines must understand and apply handgun retention techniques.

Blocking Technique

Marines perform the following blocking techniques if an opponent attempts to grab their pistol in the holster. Marines—

Place the body between the weapon and the opponent by immediately pivoting so the weapon is away from the opponent.

Step back and away from the opponent while placing the hand on the pistol grip.

Extend the left hand and block, deflect, or strike the opponent's arm.

Armbar Technique

Marines use the armbar technique when an opponent uses his right hand to grab their pistol in the holster. To execute the armbar technique, Marines—

Trap the opponent's right hand by grasping the opponent's wrist or hand with the right hand and applying pressure against the body.

Step back with the right foot and pivot sharply to the right to be next to the opponent. Always pivot in a direction that keeps the weapon away from the opponent.

Straighten the opponent's arm to apply an armbar. The arm should be straight across the torso.

Continue pivoting to the right while pulling back on the opponent's shoulder. This action may break the opponent's arm.

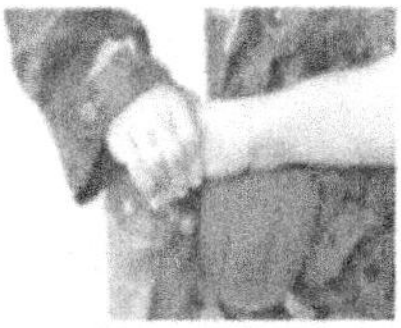

Place the left hand across the opponent's face and apply pressure back and down to take the oppo-

nent to the ground. Pressure applied beneath the nose or on the trachea is equally effective.

Note: If the opponent grabs the pistol with his left hand, Marines execute the wristlock with one hand and step in toward the opponent, rather than

away from the opponent.

Wristlock Technique

Marines use the wristlock technique when an opponent grabs their pistol while it is in the holster with his right hand. To execute the w ristlock technique, Marines—

Grasp the opponent's wrist or hand with the right hand and apply pressure against the body.

Step back and away from the opponent and pivot to the right so the weapon is away from the opponent. Always pivot in a direction that keeps the weapon away from the opponent.

Reach over the opponent's arm with the left hand and grab his hand, applying pressure against his arm with the left forearm. Execute a wristlock.

Incorporate the second hand into the wristlock and, stepping back with the left foot, pivot to the left.

Execute a two-handed wristlock by exerting downward pressure with the thumb and rotating his hand to the left.

Continue to pivot to off-balance the opponent and drive him to the ground.

Softening Techniques

Handgun retention techniques use softening techniques applied to pressure points. Bone pressure

and strikes with the hands (i.e., hammer fist), knees, and feet are also effective softening techniques.

Pressure Points. Marines use pressure point techniques to get the opponent to loosen his grip. Marines use their finger tips to apply pressure to the webbing between the index finger and thumb, the jugular notch, and the brachial plexus tie in. The following figure illustrates pressure applied to the brachial plexus tie in.

Bone Pressure. Bone pressure is the application of pressure on a bone against a hard object to initiate pain compliance. To apply bone pressure, Marines use their hand to trap the opponent's hand on the weapon. Marines apply a slow, steady pressure to the opponent's hand and fingers until his grip is softened or he releases his hold.

Strikes. If it is difficult to apply a retention technique, Marines employ strikes or kicks to force the opponent to loosen his grip. Strikes to the

eyes, the arms (radial nerve), or shoulder (brachial plexus tie in) soften the opponent's grip on the weapon.

Kicks and knee strikes to the peroneal nerve, the femoral nerve, or the groin are effective because the opponent is typically unprepared to counter the strike.

Stomping on the top of the opponent's foot may distract him or loosen his grip on the weapon.

4. Firearm Disarmament Techniques

Marines use firearm disarmament techniques during a close-range confrontation if they are un-

armed and the opponent has a firearm (pistol). These techniques are equally effective if Marines are armed but do not have time to withdraw and present the weapon. The goal of firearm disarmament techniques is to gain control of the situation

so Marines gain the tactical advantage. The goal is not necessarily to get control of the opponent's weapon.

Pistol to the Front

This technique is used when Marines are unarmed and the opponent has a pistol pointing at their front (e.g., head, chest). The technique is the same if the opponent sticks the pistol under the Marine's chin. To execute the counter when an opponent is pointing a pistol toward the front of a Marine, Marines—

Place the hands close to the weapon, about chest high, palms out.

Use the left hand to grab the opponent's forearm and push the opponent's hand with the pistol away to clear the body from in front of the weapon. At the same time, rotate the right shoulder back to clear the body from the weapon.

Maintain control of the opponent's arm.

Grasp the weapon with the right hand by placing the thumb underneath the pistol and the fingers over top of the pistol.

Keep the right hand wrapped tightly around the muzzle and quickly rotate the pistol in the opponent's hand so the muzzle is facing the opponent.

Grasp and pull the opponent's wrist or forearm away from the body while rotating the weapon.

Rotate the weapon toward the opponent while pulling it up and back and out of the opponent's grasp.

Pistol to the Rear

This technique is performed when Marines are unarmed and the opponent has a pistol pointing to the back of the Marine's head. To execute the counter when the opponent is pointing a pistol to the rear, Marines—

Place the hands close together about chest high, palms out.

Step back with the left foot, pivoting on the right foot so the side is against the opponent's front. This action clears the body from the weapon's line of fire. Keep the left hand up.

Pivot on the left foot to face the opponent and, at the same time, raise the left elbow and reach over the top of the opponent's arm with the left arm.

Wrap the left arm tightly around the opponent's arm above his elbow to control it.

Push on the opponent's shoulder with the right hand while pulling up with the left arm to achieve an armbar. This action releases the opponent's grip on the weapon. If necessary, the opponent can be taken to the ground with a leg sweep.

Note: To execute this technique, the weapon must be close to or touching the rear of the Marine. If the weapon is too far away from the body, this technique would be difficult to execute or it would be ineffective.

CHAPTER 3

HAND-HELD WEAPONS

This chapter describes all techniques for a right-handed person. However, all techniques can be executed from either side.

In drawings, the Marine is depicted in woodland camouflage utilities; the opponent is depicted without camouflage. In photographs, the Marine is depicted in woodland camouflage utilities; the opponent is depicted in desert camouflage utilities.

Marines must know how to defend against attacks when an opponent is either unarmed or armed with a held-held weapon. This chapter addresses the combative use of knives, specific weapons of opportunity, and sticks. However, virtually anything can be used as a hand-held weapon.

1. Fundamentals of Knife Fighting

Marines must be trained in knife fighting techniques. Marines experienced in offensive knife techniques can cause enough damage and massive trauma to stop an opponent. When engaged against each other, experienced knife fighters employ various maneuvers and techniques that are specific to knife fighting. Seldom, if ever, will Marines engage an opponent in a classical knife fight.

Note: When armed with a rifle, Marines are issued a bayonet. When armed with a pistol, Marine are issued a combat knife.

Angles of Attack

There are six angles from which an attack with a knife can be launched:

- Vertical strike coming straight down on an opponent.
- Forward diagonal strike coming in at a 45-degree angle to the opponent.
- Reverse diagonal strike coming in at a 45-degree angle to the opponent.
- Forward horizontal strike coming in parallel to the ground.
- Reverse horizontal strike coming in parallel to the ground.
- Forward thrust coming in a straight line to the opponent.

Target Areas of the Body

During any confrontation, the parts of the opponent's body that are exposed or readily accessible will vary. The goal in a knife fight is to attack the body's soft, vital target areas that are readily accessible (e.g, the face, the sides and front of the neck, the lower abdomen [or groin]).

Neck. Carotid arteries, located on either side of the neck, are good target areas because they are not covered by body armor or natural protection.

Lower Abdomen (or Groin). The lower abdomen (or groin region) is a good target area because it is not covered by body armor.

Heart. The heart, if not covered by body armor, is an excellent target which, if struck, can prove fatal in a matter of seconds or minutes.

Secondary Targets. There are secondary target areas that will cause substantial bleeding if an artery is severed. These target areas are not immediately fatal, but can become fatal if left unattended. Attacks to—

The legs can cause a great deal of trauma and prove fatal. For example, the femoral artery located on the inside of the thigh is a large artery which, if cut, will cause extensive blood loss.

The brachial artery, located between the biceps and triceps on the inside of the arm, can cause extensive bleeding and damage.

The arm's radial and ulnar nerves can cause extensive bleeding and damage.

Movement

Marines can move anywhere within a 360-degree circle around the opponent. This allows accessibility to different target areas of the opponent's body. Marines should avoid being directly in front of an opponent because the opponent can rely on his forward momentum to seize the tactical advantage. If Marines face an opponent, movement is made in a 45-degree angle to either side of the opponent. This angle avoids an opponent's strike and places Marines in the best position to attack an opponent.

Wearing the Combat Knife

Marines must wear the combat knife where it is easily accessible and where it can best be retained. It is recommended the combat knife be worn on the weak side hip, blade down, sharp edge facing forward. Marines can place it behind the magazine pouch where it is easily accessible to them, but not easily grabbed by an opponent.

Grip

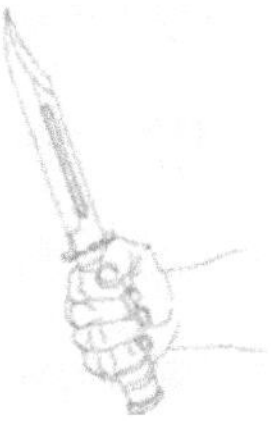

The grip on the knife should be natural. Marines grasp the knife's grip with the fingers wrapped around the grip naturally as it is pulled out of its sheath. This is commonly known as a hammer grip. The blade end of the knife is always facing the opponent.

Stance

Marines use the basic warrior stance as the foundation for knife techniques. The left hand is a vertical shield that protects either the ribs or the head and neck. The right elbow is bent with the blade pointing forward toward the opponent's head. This position serves as an index point, where all techniques are initiated.

Principles of Knife Fighting

The following are key principles of knife fighting:

Execute movements with the knife blade within a box, shoulder-width across from the neck down to the waistline. The opponent has a greater chance of blocking an attack if the blade is brought in a wide, sweeping movement to the opponent.
Close with the opponent, coming straight to the target.
Move with the knife in straight lines.
Point the knife's blade tip forward and toward the opponent.
Apply full body weight and power in each of the knife techniques. Full body weight should be put into the attack in the direction of the blade's movement (slash or thrust).
Apply constant forward pressure with the body and blade to keep the opponent off-balanced.

2. Knife Fighting Techniques

Slashing Techniques

Marines use slashing techniques to close with an enemy. Slashing techniques distract the opponent or damage the opponent so Marines can close in. Typically, Marines target the opponent's limbs, but any portion of the body that is presented can become a target.

Vertical Slash Technique. The vertical slash follows a vertical line straight down through the target. To execute the vertical slash, Marines—

Thrust the right hand out and bring the weapon straight down on the opponent, continuing to drag the knife down through the opponent's body.

Maintain contact on the opponent's body with the blade of the knife.

Forward Slash Technique. The forward slash follows a straight line in a forehand stroke, across the target areas of either the neck (high diagonal slash) or abdominal region (low horizontal slash). To execute the forward slash, Marines-
Extend the right hand while simultaneously rotating the palm up until the knife blade makes contact with the opponent.

Snap or rotate the wrist through the slashing motion to maximize blade's contact with the opponent.

Drag the knife across the opponent's body, from right to left, in a forehand stroke. The movement ends with the forearm against the body and the knife at the left hip with its blade oriented toward the opponent.

Reverse Slash Technique. The reverse slash is a follow-up technique to a forward attack. It allows Marines both a secondary attack and the ability to resume the basic warrior stance. The reverse slash follows a straight line in a backhand stroke, across the target areas of either the neck (high diagonal slash) or abdominal region (low horizontal slash). To execute the reverse slash, Marines—

Extend the right hand while simultaneously rotating the palm down until the knife blade makes contact with the opponent.

Snap or rotate the wrist through the slashing motion to maximize the blade's contact with the opponent.

Drag the knife across the opponent's body, from left to right, in a backhand stroke. Maintain contact on the opponent's body with the blade of the knife.

Thrusting Techniques

The primary objective of knife fighting is to insert the blade into an opponent to cause extensive damage and trauma. This is done with a thrusting technique. Thrusting techniques are more effective than slashing techniques because of the damage they can inflict. However, Marines use slashing techniques to close with the enemy so that they are closer to the opponent, which allows them to use the thrusting technique.

Vertical Thrust. The thrusting motion follows a vertical line straight up through the target (low into the abdomen region or high into the neck). To execute the vertical thrust, Marines—

Thrust the right hand toward the target, inserting the knife blade straight into the opponent.

Pull the knife out of the opponent.

Forward Thrust. The forward thrust follows a straight line straight into the opponent's neck (high thrust) or abdominal region (low thrust). To execute the forward thrust, Marines—

Thrust the right hand, palm down, toward the target, inserting the knife blade straight into the opponent.

Rotate the palm up once the knife is inserted to twist the blade.

Drop the right elbow and bring the knife to the opposite side of the opponent's body from where it was inserted. At the same time, rotate the hips and shoulders downward to bring body weight to bear on the attack.

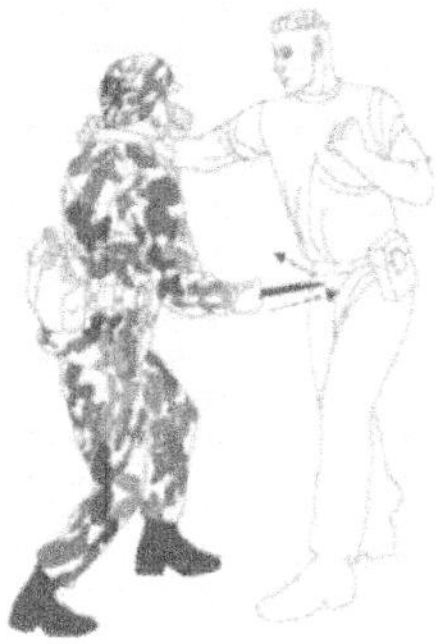

Reverse Thrust. The reverse thrust is a follow-up technique to a forward attack. It allows Marines both a secondary attack and the ability to resume the basic warrior stance. The reverse thrust follows a horizontal line straight into the opponent's neck (high thrust) or abdominal region (low thrust). To execute the reverse thrust, Marines-

Bend the right arm, crossing the arm to the left side of the body.

Thrust the right hand, palm up, toward the target, inserting the knife blade straight into the opponent.

Rotate the palm down to twist the blade once the knife is inserted.

Bring the knife to the opposite side of the opponent's body from where it was inserted.

3. Weapons of Opportunity

During an unarmed close combat situation, Marines use their bodies as weapons, but they should be ready and able to use anything around them as a weapon. For example, Marines could throw sand or liquid in an opponent's eyes to temporarily impair his vision or smash the opponent's head with a rock or helmet. Marines must use whatever means are available and do whatever it takes to take control of the situation and to win, or they face the possibility of losing their lives. Some weapons of opportunity are discussed in the following subparagraphs.

Entrenching Tool

An entrenching tool (E-tool) is commonly carried by Marines. It can be an excellent weapon, especially when sharpened. Marines can use the E-tool to block, slash, and thrust at an opponent.

Tent Pole and Pins

Marines can use tent poles and pins to block, strike, or thrust at an opponent.

Web Belt

Marines can stretch a web belt between their hands to block attacks by an opponent.

Battlefield Debris

Marines can use debris on the battlefield (e.g., sticks, glass, a sharp piece of metal) to cut, slash, or stab an opponent. They can also use other types of debris such as shovel or ax handles, boards, metal pipes, or broken rifles to strike an opponent or apply a choke.

Helmet

A helmet can be used to strike an opponent on an unprotected area like the head and face. Grasp the rim of the helmet and thrust the arms forward, striking the opponent with the top of the helmet.

4. Fundamentals of Combative Stick

On the battlefield, Marines must be ready and able to use anything as a weapon. They must learn and be able to use techniques that can be employed with most weapons of opportunity. Among these techniques are combative stick techniques. Combative stick techniques can be used with a stick, a club, a broken rifle, an E-tool, or even a web belt.

Angles of Attack

There are six angles from which an attack with a hand-held weapon can be launched:

Vertical strike coming straight down on an opponent.
Forward diagonal strike coming in at a 45-degree angle to the opponent.
Reverse diagonal strike coming in at a 45-degree angle to the opponent.
Forward horizontal strike coming in parallel to the ground.
Reverse horizontal strike coming in parallel to the ground.
Forward thrust coming in a straight line to the opponent.

Grip

Grasp the stick about 2 inches from its base.

Stance

The basic warrior stance serves as the foundation for combative stick techniques. The left hand becomes a vertical shield that protects the ribs or the head and neck. Depending on how heavy the weapon is, it should be held at a level approximately shoulder height.

Movement

Movement during combative stick techniques is the same as it is for other close combat techniques. Marines can move anywhere within a 360-degree circle around the opponent. This allows accessibility to different target areas of the opponent's body and gains a tactical advantage. Marines should avoid being directly in front of an opponent because the opponent can rely on his forward momentum to seize the tactical advantage. If Marines face an opponent, movement is made in a 45-degree angle to either side of the opponent. This angle avoids an opponent's strike and places Marines in the best position to attack an opponent.

5. Combative Stick Techniques

Strikes

Strikes are intended to inflict as much damage on an opponent as possible. Striking techniques apply to a weapon of opportunity such as a stick, a tent pole, a club, a broken rifle, an E-tool, or a Pipe-

Vertical Strike. To execute the vertical strike, Marines—

Bend the right arm, extending the weapon over the back of the right shoulder.

Rotate the forearm straight down off the elbow to bring the weapon down on the opponent.

Rotate the hips and shoulders forcefully toward the opponent.

Follow through with the strike by allowing the weight of the weapon to go through the target area of the body.

Forward Strike. To execute a forward strike, Marines—

Step forward with the left foot in the direction of the strike.

Bend the right arm with the elbow extending out to the right and the weapon extended over the right shoulder.

Rotate the forearm to the right of the elbow to bring the weapon down onto the opponent. At the same time, forcefully rotate the hips and shoulders toward the opponent.

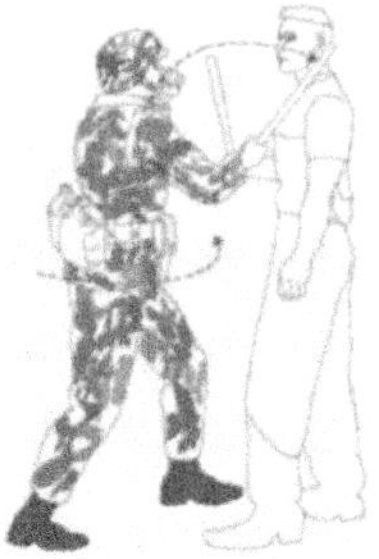

Follow through with the strike by allowing the weight of the weapon to go through the target area of the body.

Reverse Strike. The reverse strike is a follow-up technique to a forward strike. It allows Marines both a secondary attack and the ability to resume the basic warrior stance.To execute a reverse strike, Marines—

Step forward with the right foot in the direction of the strike.

Bend the right arm with the hand near the left shoulder. The weapon is extended over the left shoulder.

Rotate the forearm to the right of the elbow to bring the weapon down onto the opponent. At the same time, forcefully rotate the hips and shoulders toward the opponent.

Follow through with the strike by allowing the weight of the weapon to go through the target area of the body.

Forward Thrust. To execute the forward thrust, Marines—

Grasp the stick with the left hand, palm up, in a position where the stick can be controlled with two hands.

Lift the left leg and lunge forward off the ball of the right foot. At the same time, thrust the end of the weapon directly toward the opponent by thrusting both hands forward in a straight line.

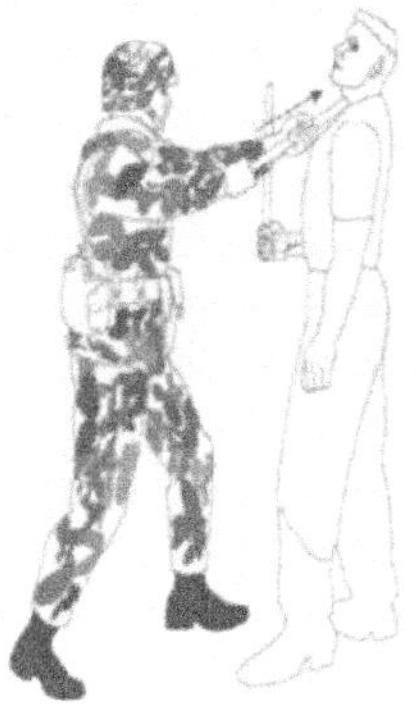

6. Blocking Techniques

A block is meant to deter or deflect an attack by an opponent. A block sets up Marines for a follow-on attack against the opponent. Blocks are executed by deflecting, rather than hitting or following through like a strike.

Blocks Against Unarmed Attacks

To block against an unarmed attack, Marines—

Step forward at a 45-degree angle with either the right or left foot. This moves the body out of the line of attack.

Raise the left arm and block the strike with the meaty portion of the forearm.

Employ the stick by one of two techniques:

Strike the opponent with the stick.
Use the stick to block on two points of contact. When the stick is used to block, it serves as an incidental strike. This technique is only used if Marines have closed with the opponent and are inside his strike.

Blocks Against Armed Attacks

Block for a Vertical Strike. To execute the block for a vertical strike, Marines—

Step forward with the left foot at a 45-degree angle to the left. This moves the body out of the line of attack.

Block on two points of contact to disperse the impact of the attack:

Block the opponent's stick by positioning the stick so it is perpendicular to the opponent's stick. If the stick is not perpendicular to the opponent's stick, the stick can slide through and make contact on the Marine.
Block the opponent's wrist or forearm with the back of the left forearm.

Note: Use the stick to block the opponent's arm if closer to the opponent. It is the same movement as blocking with the arm, except the opponent's arm is blocked with both the stick and the arm.

Block for a Forward Strike. To execute the block for a forward strike, Marines-
Step forward with the left foot at a 45-degree angle to the left. This moves the body out of the line of attack and inside the opponent's strike.

Block on two points of contact to disperse the impact of the attack:

- Block the opponent's wrist or forearm with the meaty portion of the back of the left forearm.
- Strike the opponent's attacking biceps with the stick.

Block for a Reverse Strike. To execute the block for a reverse strike, Marines—

Step forward at a 45-degree angle with the right foot to the right. This moves the body out of the line of attack and inside the opponent's strike.

Block on two points of contact to disperse the impact of the attack:

Block the opponent's stick by positioning the stick so it is perpendicular to the opponent's stick.

Block the opponent's forearm with the meaty portion of the left forearm.

Note: If closer to the opponent, block his triceps with the back of the left forearm and strike his forearm with the stick.

7. Unarmed Against Hand-Held Weapons

If Marines are engaged against an opponent with a knife, a stick, or some other weapon of opportunity, they must establish and maintain an offensive mindset, not a defensive mindset. Their survival depends on it. They cannot afford to think about getting cut or hurt.

Angles of Attack

Before Marines can learn to block or counter an attack with a hand-held weapon (e.g., knife, stick), they must know from what angle the opponent is attacking. There are six angles from which an opponent will typically attack with a hand-held weapon:

Vertical strike coming in straight down on the Marine.
Forward diagonal strike coming in at a 45-degree angle to the Marine.
Reverse diagonal strike coming in at a 45-degree angle to the Marine.
Forward horizontal strike coming in parallel to the ground.
Reverse horizontal strike coming in parallel to the ground.
Forward thrust coming in a straight line to the Marine.

Blocks

If the opponent has a hand-held weapon (e.g., knife or stick), Marines parry the opponent's hand or arm to block the attack.

Basic Block. To execute the basic block technique, Marines-

Step forward at a 45-degree angle to move out of the line of the attack. Always step in the direction of the strike.

Thrust the forearms forward, hands up, against the opponent's attacking arm. Contact is made on the opponent's arm with the backs of the forearms.

Block for a Vertical Strike. To execute the block against a vertical strike, Marines—

Step forward with the left foot at a 45-degree angle to the left to move out of the line of attack.

Thrust the forearms forward, hands up, against the outside of the opponent's attacking arm.

Block for a Forward Strike. To execute the block against a forward diagonal or forward horizontal strike, Marines-

Step forward with the left fool inside the opponent's attacking arm.

Block the attack with both arms bent so the forearms make contact with the opponent's biceps and forearm.

Block for a Reverse Strike. To execute the block against a reverse diagonal or reverse horizontal strike, Marines—

Step forward with the right foot to the outside of the opponent's attacking arm.

Block the attack with both arms bent so the forearms make contact with the opponent's triceps and forearm.

Block for a Forward Thrust. To execute the block against a forward thrust, Marines—

Bend at the waist, move the hips backward, and jump backward with both feet to move away from the attack. This action is known as "hollowing out."

Hollow out and block the attack with the arms bent and hands together on top of the attacking arm.

Overlap the hands slightly so one thumb is on top of the other hand's index finger. The other thumb should be under the other hand's index finger.

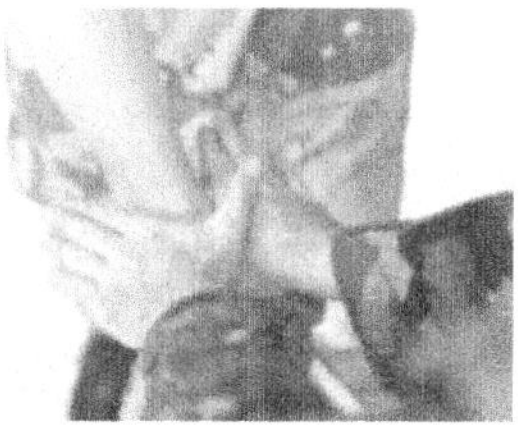

8. Counters to Hand-Held Weapon Attacks

Principles of Counters

A counter is used to control the situation to regain the tactical advantage and end the fight. Regardless of the type of weapon or angle of attack, the following principles apply to countering the attack with a hand-held weapon:

Move out of the line of attack. Movement is executed in a 45-degree angle forward to the left or right.

- Block the attack.

Note: The first two actions are taken simultaneously.

- Take control of the weapon by controlling the hand or arm that is holding the weapon. Never attempt to grab the opponent's weapon.
- Execute the appropriate follow-up to end the fight; e.g., strikes, joint manipulations, throws, or takedowns (see chap. 8). Marines should continue to attack the opponent until the fight ends.

Counter Techniques

There are two techniques that can be used to counter any armed attack: forward armbar counter and reverse armbar counter. These techniques can be used to counter a vertical attack, a forward diagonal strike, or a forward horizontal strike. With minor variations, the same techniques are used to counter reverse strikes. A third technique, the bent armbar counter, is used to counter a vertical attack.

Forward Armbar Counter. To execute the forward armbar counter to an attack coming from a forward strike, Marines-

Step forward with the left foot inside the opponent's attacking arm.

Block the attack with both arms bent so the forearms make contact with the opponent's biceps and forearm.

Slide the left arm over the opponent's forearm and wrap the arm tightly around his arm, trapping the opponent's attacking arm between the biceps and torso.

Place the right hand on the opponent's shoulder or upper arm to further control his arm and to effect an armbar.

Execute an armbar and continue to exert steady pressure against the arm.

Reverse Armbar Counter. To execute the reverse armbar counter to an attack coming from a forward strike, Marines-

Step forward with the left foot inside the opponent' s attacking arm.

Block the attack with both arms bent so the forearms make contact with the opponent's biceps and forearm.

Control the opponent's arm with the left hand and pivot to the right so the back is against the opponent's side. Immediately slide the right arm over the opponent's biceps and wrap the arm tightly around his arm, trapping the opponent's attacking arm between the biceps and torso.

Grasp the opponent's wrist with the left hand and twist his thumb away from the body.

Take the opponent to the ground with an armbar takedown if he does not drop the weapon from the wristlock.

Bent Armbar Counter. This counter is particularly effective against a vertical attack. To execute the bent armbar counter, Marines—

Step forward with the left foot inside the opponent's attacking arm.

Block the attack with both arms bent so the forearms make contact with the opponent's biceps and forearm.

Grasp the opponent's forearm with the left hand. At the same time, slide the right arm underneath the opponent's triceps and grasp the opponent's forearm or wrist with the right hand.

Apply downward pressure with the hands against the opponent's forearm to off-balance the opponent.

(reverse blank)

CHAPTER 4

STRIKES

This chapter describes all techniques for a right-lmnded person. However, all techniques can be executed from either side.

In drawings, the Marine is depicted in woodland camouflage utilities; the opponent is depicted without camouflage. In photographs, the Marine is depicted in woodland camouflage utilities; the opponent is depicted in desert camouflage utilities.

Strikes are unarmed individual hitting techniques. Strikes use the hands, elbows, knees, feet, and, in some instances, other parts of the body as personal weapons. Marines must know how to execute strikes effectively. They must also know how to counter strikes from an opponent.

1. Principles of Punches

Muscle Relaxation

Muscle relaxation is crucial when executing punches. The natural tendency in a fight is to tense up, which results in rapid fatigue and decreased power generation. Marines who remain relaxed during a close combat situation generate greater speed, which results in greater generation of power. Relaxing the forearms generates speed and improves reaction time. At the point of impact, Marines clench the fist to cause damage to the opponent and avoid injury to the wrist and hand.

Weight Transfer

Weight transfer is necessary to generate power in a punch. Marines accomplish this by-

Rotating their hips and shoulders into the attack.

Moving their body mass straight forward or backward in a straight line.

Dropping their body weight into an opponent. Body mass can be transferred into an attack from high to low or from low to high.

Rapid Retraction

When Marines deliver a punch, rapid retraction of the fist is important. Once the hand has made contact with the target, Marines quickly return to the basic warrior stance. Rapid retraction-

Returns the hand and arm to the protection afforded by the basic warrior stance.

Prevents the opponent from being able to grab the hand or arm.

Permits the hand and arm to be "chambered" or "re-cocked" in preparation for delivering a subsequent punch.

Telegraphing

Telegraphing a strike occurs when body movements inform the opponent of the intention to launch an attack. Staying relaxed helps to reduce telegraphing.

Often, an untrained fighter telegraphs his intention to attack by drawing his hand back in view of his opponent, changing facial expression, tensing neck muscles, or twitching. These movements, however small, immediately indicate an attack is about to be delivered. If the opponent is a trained fighter, he may be able to evade or counter the attack. If the opponent is an untrained fighter, he may still be able to minimize the effect of an attack.

2. Punches

Punches may be thrown during any hand-to-hand confrontation. Most people resort to punching because it is a natural reaction to a threat. The purpose of a punch is to stun the opponent or to set him up for a follow-up finishing technique. However, punches should only be executed to the soft tissue areas of an opponent. A correctly delivered punch maximizes the damage to an opponent while minimizing the risk of injury to Marines.

Basic Fist

Punches are executed using the basic fist. To make the basic fist, the fingers are curled naturally into the palm of the hand and the thumb is placed across the index and middle fingers. Do not clench the fist until movement has begun. This reduces muscular tension in the forearm and increases speed and reaction time. Just before impact, Marines exert muscular tension on the hand and forearm to maximize damage to the opponent and reduce their chances of injury. Contact should be made with the knuckles of the index and middle fingers.

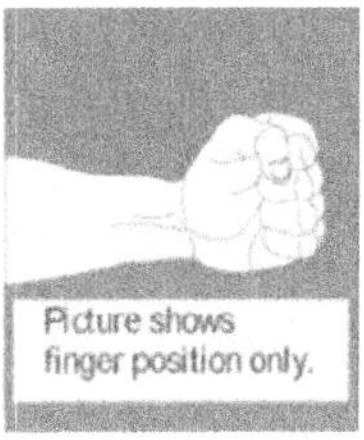

When striking with the basic fist, Marines must keep the hand straight, or in line, with the wrist to avoid injury to the wrist.

Lead Hand Punch

The lead hand punch is a snapping straight punch executed by the forward or lead hand. It is a fast punch designed to keep the opponent away and to set up an attack. A lead hand punch conceals movement and allows Marines to get close to the opponent. Lead hand punches should strike soft tissue areas, if possible. To execute the lead hand punch, Marines-

Snap the lead hand out to nearly full extension, while rotating the palm to the ground.

Contact the opponent with the first two knuckles of the fist.

Retract the hand immediately, resuming the basic warrior stance.

Rear Hand Punch

The rear hand punch is a snapping punch executed by the rear (right) hand. It is a power punch designed to inflict maximum damage on the opponent. Its power comes from pushing off the rear leg and rotating the hips and shoulders. To execute the rear hand punch, Marines—

Rotate the hips and shoulders forcefully toward the opponent and thrust the rear hand straight out, rotating the palm down, to nearly full extension.

Shift body weight to the lead foot while pushing off on the ball of the rear foot.

Contact the opponent with the first two knuckles of the fist.

Retract the hand immediately.

Uppercut

The uppercut is a powerful punch originating below the opponent's line of vision. It is executed in an upward motion traveling up the centerline of the opponent's body. It is delivered in close and usually follows a preparatory strike that leaves the target area unprotected. When delivered to the chin or jaw, the uppercut can render an opponent unconscious, cause extensive damage to the neck, or sever the tongue. To execute the uppercut, Marines—

Bend the arms, rotating the palm inboard. The distance the arms bend depends on how close the opponent is.

Rotate the hips and shoulders forcefully toward the opponent, thrusting the fist straight up toward the opponent's chin or jaw.

Contact the opponent with the first two knuckles of the fist.

Retract the hand immediately.

Hook

The hook is a powerful punch that is executed close in and is usually preceded by a preparatory strike. To execute the hook, Marines—

Thrust the right arm in a hooking motion toward the opponent, keeping the elbow bent while forcefully rotating the right shoulder and hip toward the opponent.

Contact the opponent with the first two knuckles of the fist. Continue rotating the shoulder and hip, following through with the fist to the target.

Retract the hand immediately.

3. Strikes with the Upper Body

Strikes stun the opponent or set him up for a follow-up finishing technique. The hands, forearms, and elbows are individual weapons of the arms that can be used to execute strikes including the hammer fist, chin jab, knife hand, eye gouge, and elbow strikes. These strikes provide a variety of techniques that can be used in any type of close combat situation.

Principles of Execution

There are several principles of execution that ensure a strike's effectiveness.

Generating Power. Weight transfer is necessary to generate maximum power in a punch. Marines accomplish this by-

Rotating the hips and shoulders into the attack.
Moving the body mass straight forward or backward in a straight line.
Dropping body weight into an opponent. Body mass can be transferred into an attack from high to low or from low to high.

Muscular Tension. The arms are relaxed until the moment of impact. At the point of impact, Marines apply muscular tension in the hand and forearm to maximize damage to the opponent and to avoid injury to the hand. The arms are relaxed until the moment of impact.

Hit and Stick and Follow-Through. A strike should be delivered so that the weapon (e.g., hand, elbow) hits and remains on the impact site (target) and follows through the target. This technique inflicts maximum damage on the opponent.

Strikes with the arms are executed with "heavy hands"; i.e., the strike is executed by driving through with the strike to allow the weight of the hand to go through the target area of the body. Contact with the opponent should be made with the arm slightly bent, and the arm extends as it moves through the target. This technique allows Marines to deliver strikes effectively without executing full force.

Movement. Movement puts Marines in the proper position for launching an attack against an opponent as well as providing protection.

Movement is initiated from the basic warrior stance and ends by resuming the basic warrior stance. Strikes can be performed with either the left or right arm depending upon—

The angle of attack.
The position of the opponent.
The opponent's available, vulnerable target areas.

Target Areas of the Body

For each strike, there are target areas of the body that, when struck, maximize damage to an opponent. Strikes use gross motor skills as opposed to fine motor skills. The target areas of the body are just that—areas. Pinpoint accuracy on a specific nerve is not needed for the strike to be effective.

Hammer Fist

Striking with the hammer fist concentrates power in a small part of the hand, which, when transferred to the target, can have a devastating effect. The striking surface of the hammer fist is the meaty portion of the hand below the little finger. To execute the hammer fist strike, Marines-

Make a fist and bend the arm at approximately a 45- to 90-degree angle. At the same time, rotate the right hip and right shoulder backward.

Thrust the fist forward onto the opponent while rotating the right hip and shoulder forward.

Rotate the wrist so the hammer fist makes contact on the opponent.

Chin Jab

The chin jab can immediately render an opponent unconscious and cause extensive damage to the neck and spine. The striking surface is the heel of the palm of the hand. To execute the chin jab, Marines—

Bend the right wrist back at a 90-degree angle with the palm facing the opponent and the fingers pointing up.

Keep the right arm bent and close to the body. Extend the hand into a concave position with the fingers slightly spread apart.

Step forward with the left foot toward the opponent, keeping the feet approximately shoulder-width apart and the knees bent. This is done to close with the opponent.

Keep the right arm bent and close to the side. Thrust the palm of the hand directly up under the opponent's chin. At the same time, rotate the right hip forward to drive the body weight into the attack to increase the power of the strike. The attack should travel up the centerline of the opponent's chest to his chin.

Knife Hand

The knife hand is one of the most versatile and devastating strikes. The striking surface is the cutting edge of the hand, which is the meaty portion of the hand below the little finger extending to the top of the wrist. The striking surface is narrow, allowing strikes on the neck between the opponent's body armor and helmet. The knife hand strike is executed from one of three angles: outside, inside, and vertical.

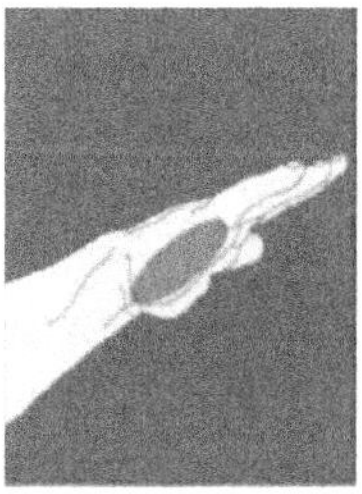

Outside Knife Hand. To execute the outside knife hand strike, Marines—

Execute a knife hand by extending and joining the fingers of the right hand and placing the thumb next to the forefinger (like saluting).

Retract the right hand. At the same time, rotate the right hip and right shoulder backward.

Thrust the knife hand forward (horizontally) onto the opponent while rotating the right hip and shoulder forward.

Inside Knife Hand. To execute the inside knife hand strike, Marines—

Execute a knife hand.

Bring the right hand over the left shoulder. At the same time, rotate the right shoulder forward and the left hip forward.

Thrust the knife hand forward (horizontally) onto the opponent while rotating the right hip and shoulder forward and the left shoulder backward.

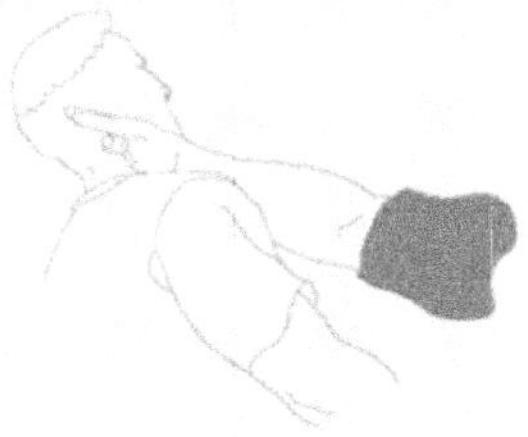

Vertical Knife Hand. When thrown vertically, the knife hand strike comes straight down in a straight line.

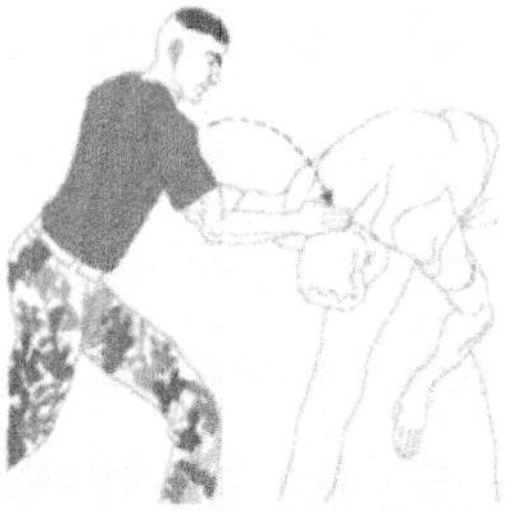

Eye Gouge

The eye gouge is used to attack an opponent's eyes, blinding him so follow-up strikes can be executed. The striking surface is the tips of the fingers or thumb. To execute the eye gouge, Marines—

Extend the right hand with the fingers slightly spread apart to allow entry into the eye sockets.

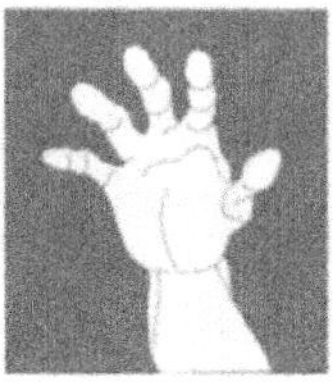

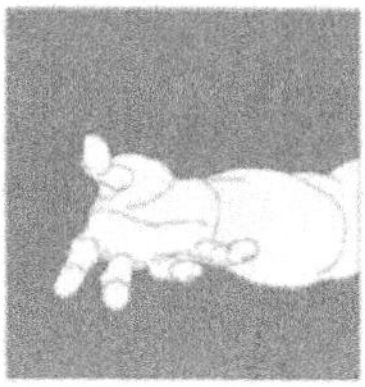

Place the palm of the hand either toward the ground or toward the sky and thrust the right hand forward into the opponent's eyes.

Thrust the hand forward at the opponent's nose level so the fingers or thumb slide naturally into the grooves of the opponent's eye sockets.

Elbow Strikes

The elbow is a powerful weapon that can be used in several different ways to attack virtually any part of an opponent's body. Elbow strikes can be performed either vertically (upward or downward) or horizontally (forward or reverse). The striking surface is 2 inches above or below the point of the elbow, depending upon the angle of attack, the opponent's attack angle, and the position of the opponent.

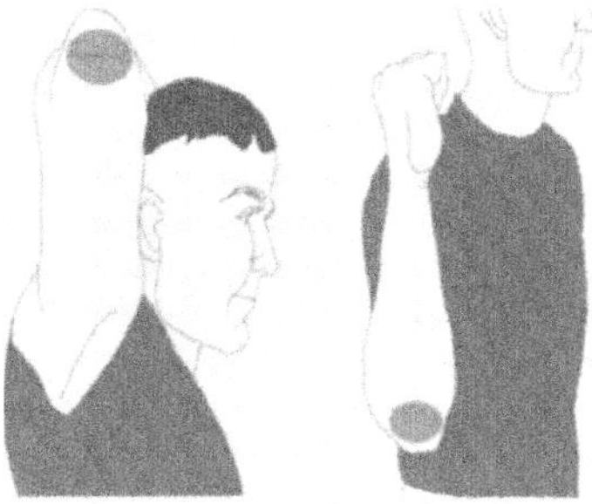

Above Elbow **Below Elbow**

Vertical Elbow Strike (Up). To execute an upward vertical elbow strike, Marines-

Bend the right elbow, keeping the fist close to the body. The fist is at shoulder level and the elbow is next to the torso.

Thrust the elbow vertically upward toward the opponent while rotating the right shoulder and hip forward to generate additional power.

Contact the opponent with the right forearm 2 inches above the point of the elbow.

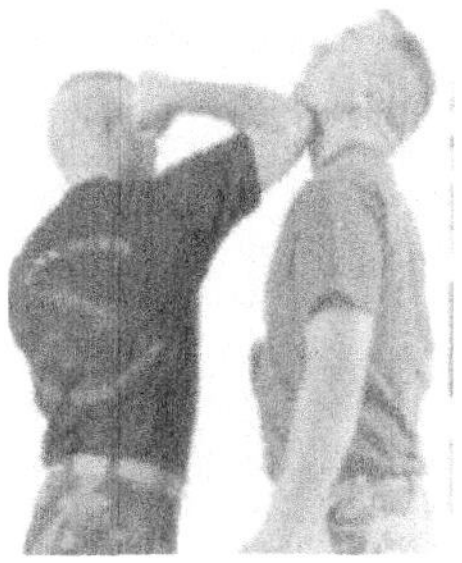

Vertical Elbow Strike (Down). To execute a downward vertical elbow strike, Marines—

Bend the right elbow, keeping the fist close to the body. The fist is on the shoulder and the elbow is raised well above the shoulder.

Thrust the elbow vertically downward toward the opponent while dropping body weight into the attack to generate additional power.

Contact the opponent with the right triceps 2 inches above the point of the elbow.

Horizontal Elbow Strike (Forward). To execute the forward horizontal elbow strike, Marines-

Tuck the right fist near the chest with the palm heel facing the ground.

Thrust the right elbow horizontally forward toward the opponent. The forearm is parallel to the ground.

Rotate the right shoulder and hip forward to generate additional power.

Contact the opponent with the right forearm 2 inches below the point of the elbow.

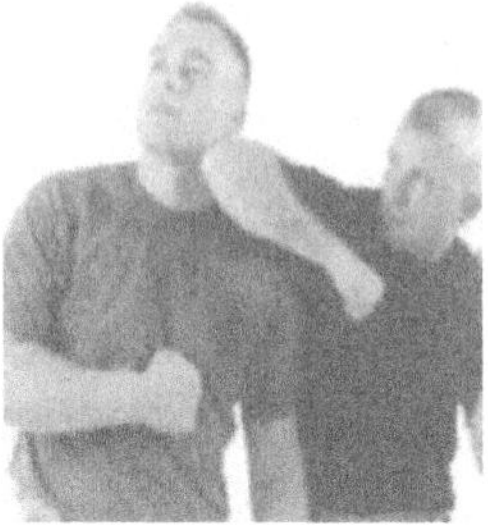

Horizontal Elbow Strike (Rear). To execute the rear horizontal elbow strike, Marines-

Tuck the right fist near the left shoulder with the palm heel facing the ground. At the same time, rotate the right shoulder forward and the left hip forward.

Thrust the right elbow horizontally rearward toward the opponent. The forearm is parallel to the ground and the hand moves toward the direction of the attack.

Rotate the right hip back and the right shoulder backward to generate additional power.

Contact the opponent with the right triceps 2 inches above the point of the elbow.

4. Strikes with the Lower Body

The legs are the body's most powerful weapons because they use the largest muscle groups to generate a strike. Legs are also less prone to injury. The feet are the preferred choice for striking because they are protected by boots. Marines use their feet, heels, and knees to execute kicks, knee strikes, and stomps.

Kicks

The purpose of a kick is to stop an opponent's attack or to create an opening in his defense in order to launch an attack. Kicks can be performed with the left (lead) leg or the right (rear) leg. Kicks with the rear leg have greater power because the hips are rotated into the attack. However, the rear leg is further away from the opponent so a strike with the rear leg will not contact the opponent as quickly as a strike with the lead leg.

Front Kick. The front kick is executed when the opponent is in front of the Marine. The front kick, delivered with either the rear or lead leg, is effective for striking below the waist. Attempting to kick higher results in diminished balance and provides the enemy with a greater opportunity to grab the leg or foot. The striking surfaces arc the toe of the boot or the bootlaces. To execute the front kick, Marines-

Raise the left knee waist high, pivot the hips into the attack, and thrust the left foot forward toward the opponent.

Contact the opponent with the toe of the left boot or bootlaces.

Return to the basic warrior stance.

Side Kick. The side kick, delivered with the lead leg, is effective for striking the knees. The side kick is executed when the opponent is to the side of the Marine. The striking surface is the outside cutting edge of the boot near the heel. To execute the side kick, Marines—

Raise the right knee waist high and rotate the right hip forward.

Thrust the right foot to the right side toward the opponent, turning the foot at a 90-degree angle to maximize the striking surface on the opponent.

Contact the opponent with the cutting edge of the right boot.

Return to the basic warrior stance.

Knee Strikes

Knee strikes are excellent weapons during the close range of close combat fighting. The knee strike is generally delivered in close.

Vertical Knee Strike. The striking surface is the thigh, slightly above the knee. To execute the vertical knee strike, Marines-

Raise the right knee and drive it up forcefully into the opponent. Power is generated by thrusting the leg upward.

Contact the opponent 2 inches above the right knee.

Horizontal Knee Strike. The horizontal knee strike is executed with the leg generally parallel with the ground while rotating the hips to generate power. It is often delivered to the peroneal nerve. The striking surface is the front of the leg, slightly above or below the knee. To execute the horizontal knee strike, Marines-

Raise the right knee, rotate the right hip forward while pivoting on the left foot, and drive the knee horizontally into the opponent.

Contact the opponent 2 inches above the right knee.

Stomps

Stomps are delivered with the feet, usually when the opponent is down. Remember, when the opponent is down, Marines take whatever target is available.

Vertical Stomp. The vertical stomp allows Marines to remain upright and balanced, to rapidly deliver multiple blows with either foot, and to quickly and accurately attack the target. It is the preferred stomp. The striking surface is the flat bottom of the boot or the cutting edge of the heel. To execute the vertical stomp, Marines—

Raise the right knee above the waist with the right leg bent at approximately a 90-degree angle.

Drive the flat bottom of the right boot or the cutting edge of the right heel down onto the opponent forcefully. At the same time, bend the left knee slightly to drop the body weight into the strike.

Ax Stomp. The striking surface of the ax stomp is the cutting edge of the heel. To execute the ax stomp, Marines-

Raise the right heel above the waist, keeping the right leg straight.

Drive the cutting edge of the right heel down onto the opponent forcefully. At the same time, bend the left knee slightly to drop the body weight into the strike.

5. Counters to Strikes

In a close combat situation, an opponent will attempt to strike Marines with punches and kicks. When an opponent uses a strike, Marines must first avoid the strike. This is accomplished by moving quickly and blocking. Next, Marines must get into an offensive position. This allows Marines to use offensive strikes to attack the opponent. Regardless of the strike, the counter to a strike requires Marines to move, block, and strike.

Move

The first step in countering a strike is to move out of the way of the strike's impact. Movement removes Marines from the opponent's intended strike point and positions Marines to attack. Movement is executed at approximately a 45-degree angle to the front or rear. Movement is always initiated from the basic warrior stance. Return to the basic warrior stance with the toe of the lead foot pointing toward the opponent once the movement is complete.

Block

Different blocks are executed based on the strike. These will be covered by individual counters.

Strike

Any of the upper body or lower body strikes can be executed as a follow-on attack or part of the counter to an opponent's strike. The follow-on strike is determined by the angle to the opponent, the position of the opponent, and the opponent's available vulnerable target areas.

Counters to Punches

Counter to a Lead Hand Punch. This counter is used when the opponent throws a lead hand punch. To execute the counter to the lead hand punch, Marines-

Step forward and to the left at approximately a 45-degree angle, moving in to the outside of the opponent's attacking arm.

Raise the left arm and block or deflect the opponent's lead hand with the palm of the hand or the meaty portion of the forearm.

Hit and stick by leaving the left arm against the opponent's right arm while stepping forward and to the left at approximately a 45-degree angle to close with the opponent.

Execute a finishing technique, such as a strike or a kick, to the opponent's exposed target areas.

Counter to a Rear Hand Punch. This counter is used when the opponent throws a rear hand punch. To execute the counter to the rear hand punch, Marines—

Step forward and to the left at approximately a 45-degree angle, moving in to the outside of the opponent's attacking arm.

Raise the left arm and block or deflect the opponent's rear hand with the palm of the hand or the meaty portion of the forearm.

Hit and stick by leaving the left arm against the opponent's right arm while stepping forward and to the right at approximately a 45-degree angle to close with the opponent.

Execute a finishing technique, such as a strike or a kick, to the opponent's exposed target areas.

Counters to Kicks

Counter to a Front Kick (Left or Lead Leg). This counter is used when the opponent executes a front kick with his left leg. To execute the counter to a front kick, Marines-

Step forward and to the right at approximately a 45-degree angle, moving in to the outside of the opponent's striking leg.

Lower the left arm and block or deflect the opponent's leg with the palm of the hand or the meaty portion of the forearm.

Hit and stick by leaving the left arm against the opponent's leg while stepping forward and to the left at approximately a 45-degree angle to close with the opponent.

Execute a finishing technique, such as a strike or a kick, to the opponent's exposed target areas.

Counter to a Front Kick (Right or Rear Leg). This counter is used when the opponent executes a front kick with his right leg. To execute the counter to a front kick, Marines—

Step forward and to the left at approximately a 45-degree angle, moving in to the outside of the opponent's striking leg.

Lower the left arm and block or deflect the opponent's leg with the palm of the hand or the meaty portion of the forearm.

Hit and stick by leaving the left arm against the opponent's leg while stepping forward and to the right at approximately a 45-degree angle to close with the opponent.

CHAPTER 5

THROWS

This chapter describes all techniques for a right-handed person. However, all techniques can be executed from either side.

The Marine is depicted in camouflage utilities. The opponent is depicted without camouflage.

Marines use throwing techniques to maintain the tactical advantage and to throw the opponent to the ground during close combat. Throws apply the principles of balance, leverage, timing, and body position to upset an opponent's balance and to gain control by forcing the opponent to the ground. Throwing techniques are effective because they are size- and gender-neutral, and they rely on the momentum and power generated by the opponent rather than the strength or size of the Marine. Marines also execute a throw as a devastating attack against an opponent, possibly causing unconsciousness or broken limbs. When Marines execute throws, they must maintain balance and, simultaneously, prevent the opponent from countering a throw or escaping after being forced to the ground.

1. Turning Throw

Marines use a turning throw to take the opponent to the ground while they remain standing. A turning throw can also be executed from a stationary position. It is particularly effective if Marines and the opponent are wearing gear. To execute the turning throw, Marines-

Grasp the opponent's right wrist with the left hand.

Step forward with the right foot, place it against the outside of the opponent's right foot, and pivot so the back of the heel is next to the middle of the opponent's foot.

Hook the opponent's right arm with the right arm and pinch his arm between the biceps and forearm, touching the opponent with the body.

Pull the opponent's wrist downward, keeping it close to the body.

Pivot to the left on the ball of the foot and continue pulling downward on the opponent's wrist while rotating the wrist outward to off-balance him.

2. Hip Throw

Marines use a hip throw to take the opponent to the ground while they remain standing. A hip throw is particularly effective if the opponent is moving forward or pushing. Marines use the opponent' s forward momentum to execute the hip throw. To execute the hip throw, Marines-

Grasp the opponent's right wrist with the left hand.

Step forward with the right foot and place it against the outside of the opponent's right foot.

The back of the heel should be next to the opponent' s foot.

Step back with the left foot and rotate on the ball of the foot. The back of the heel is next to the opponent' s toe. The knees are bent.

Rotate at the waist and hook the right hand around the back of the opponent's body (anywhere from his waist to his head). The side and hip should be against the opponent.

Rotate the hip against the opponent. The hips must be lower than the opponent's.

Use the right hand to pull the opponent up on the hip to maximize contact.

Pull the opponent's arm across the body and, at the same time, lift the opponent off the ground slightly by bending at the waist, straightening the legs, and rotating the body to the left.

3. Leg Sweep

Marines use a leg sweep to take the opponent to the ground while they remain standing. A leg sweep is particularly effective if the opponent is already off-balanced and moving backward or pulling on the Marine. To execute the leg sweep, Marines—

Grab the opponent's right wrist with the left hand and grab the opponent' s left shoulder with the right hand.

Note: Grab the opponent's clothing or gear if his wrist and shoulder cannot be grabbed.

Step forward with the left foot and place it on the outside of the opponent's right foot.

Pull the opponent's wrist downward, close to the body, and push his shoulder backward to off-balance him.

Raise the right knee no higher than the waist.

Kick the foot past the opponent's right leg.

Use the heel of the boot to make contact with the opponent's calf (anywhere from the top of the calf down to the Achilles tendon or on the inside of the calf).

Sweep the opponent to the ground.

CHAPTER 6

CHOKES AND HOLDS

This chapter describes all techniques for a right-handed person. However, all techniques can be executed from either side.

The Marine is depicted in camouflage utilities. The opponent is depicted without camouflage.

WARNING

During training, never execute a choke at full force or full speed and never hold a choke for more than 5 seconds.

When Marines correctly perform a choke, they render an opponent unconscious in as little as 8 to 13 seconds. Chokes are easily performed regardless of size or gender. Marines must know how to apply chokes and how to counter a choke or a hold executed by an opponent.

1. Types of Chokes

There are two types of chokes: an air choke and a blood choke. An air choke closes off the airway to the lungs, thereby preventing oxygen from reaching the heart. A blood choke cuts off the blood flow to the brain. Both types can result in unconsciousness and eventual death for an opponent.

Air Choke

An air choke is performed on the opponent's windpipe (or trachea), cutting off the air to the lungs and heart. If Marines execute the air choke properly, the opponent loses consciousness within 2 to 3 minutes. Due to the length of time it takes to immobilize the opponent, air chokes are not recommended.

Blood Choke

A blood choke is performed on the opponent's carotid artery, which carries oxygen-enriched blood from the heart to the brain. The carotid artery is located on both sides of the neck. If Marines execute a blood choke properly, the opponent will lose consciousness within 8 to 13 seconds. The blood choke is the preferred choke because its intended effect (i.e., the opponent losing consciousness) can be executed quickly, ending the fight.

2. Chokes

Front Choke

Marines execute a front choke when they are facing the opponent. Marines use the opponent's lapels or collar to execute a front choke. To execute the front choke, Marines-

Grab the opponent's right lapel with the right hand, making certain that the knuckles or the back of the hand are against the opponent's carotid artery on the right side of his neck.

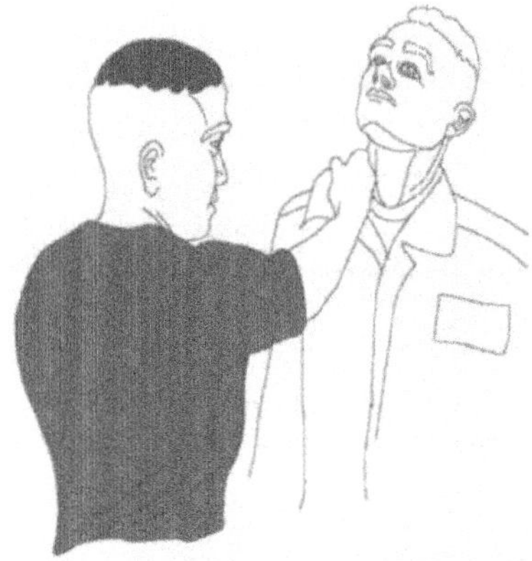

Keep the right hand against the opponent's neck, reach under the opponent's right arm with the left hand, grab the opponent's left lapel, and form an X with the wrists.

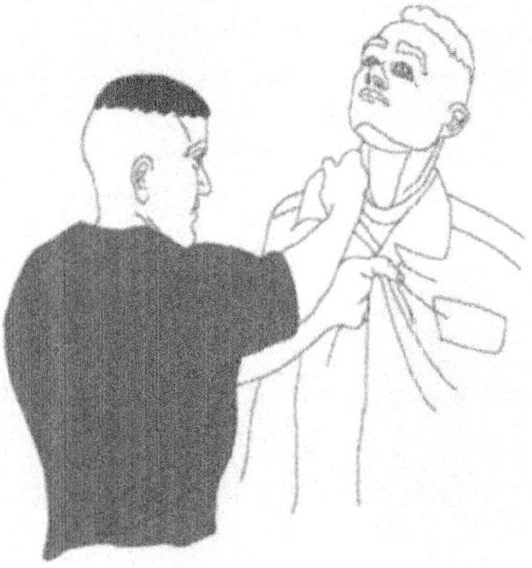

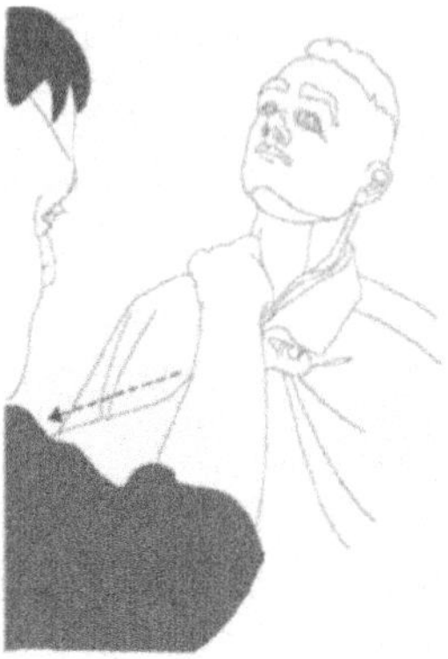

Press the right hand against the opponent's carotid artery.

Pull his lapel to the left with the left hand.

Side Choke

Marines execute a side choke when they are facing the opponent. The side choke is particularly effective when deflecting a punch thrown by an opponent. To execute the side choke, Marines—

Use the left hand to parry the opponent's arm inboard (to the inside of the opponent's reach).

Bring the right arm underneath the opponent's arm and up around the front of his neck.

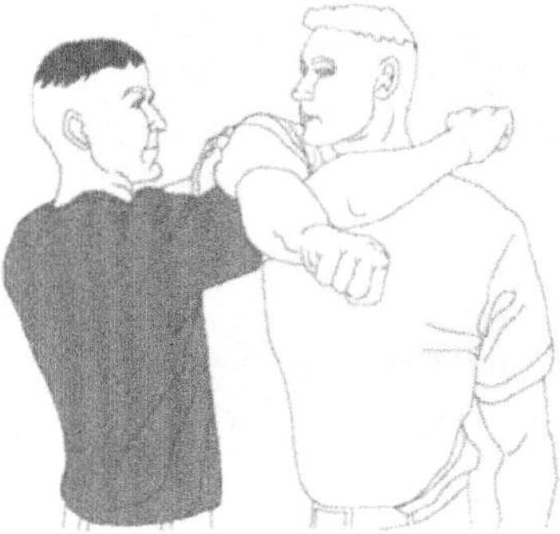

Extend the fingers, place the back of the forearm against the opponent's neck just below his ear, and press the carotid artery on his neck.

Note: If Marines are unable to place their fingers against the opponent's neck, they may use the back of the thumb or wrist.

Reach, with the left hand, around the back of the opponent's neck and clasp the hands together.

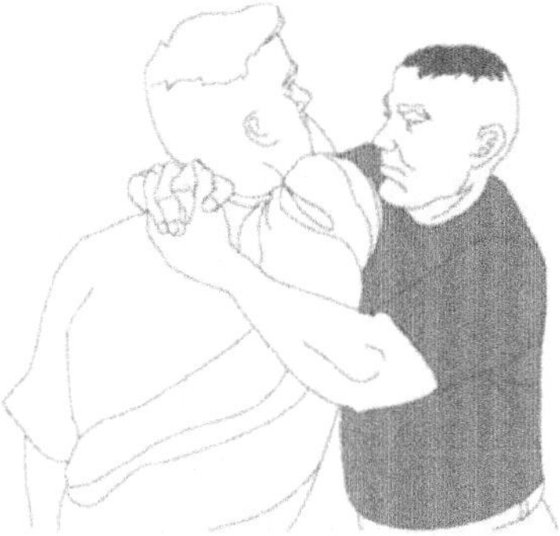

Pull the opponent toward the chest by pulling the clasped hands toward the chest.

Exert pressure on the side of the opponent's neck with the forearm.

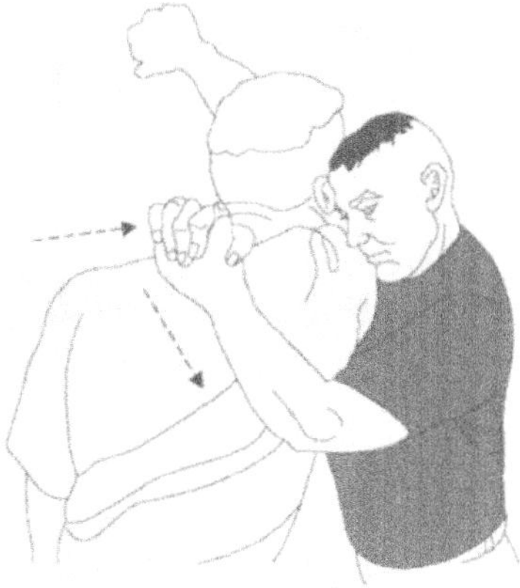

Rear Choke

Marines execute a rear choke when they are behind the opponent, the opponent is on the ground, or when they are taking the opponent to the ground. To execute the rear choke, Marines—

Reach, with the right arm, over the opponent's right shoulder and hook the bend of the arm around his neck.

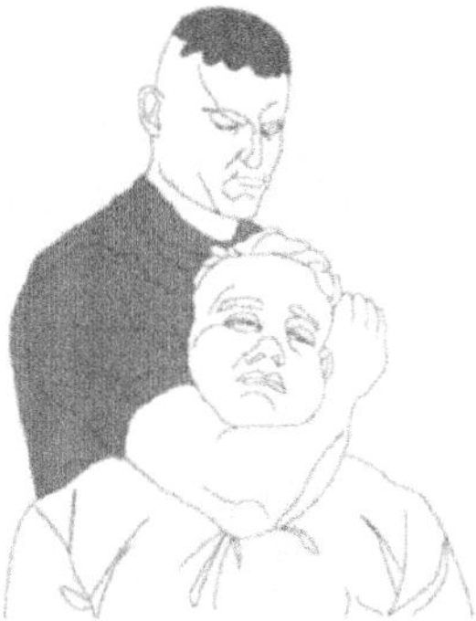

Clasp both hands together.

Exert pressure with the biceps and forearms on both sides of the opponent's neck on his carotid arteries.

Maintain pressure with the biceps and forearms on both sides of the neck and draw the opponent closer by drawing the right arm in.

Figure-Four Choke

A variation of the rear choke is the figure-four choke. The figure-four choke allows Marines to gain more leverage than the rear choke. If the rear choke cannot be secured, the figure-four variation is employed to increase the pressure of the choke on the opponent. To execute the figure-four variation of the rear choke, Marines—

Apply a rear choke. The Marine's body should be against the opponent's body.

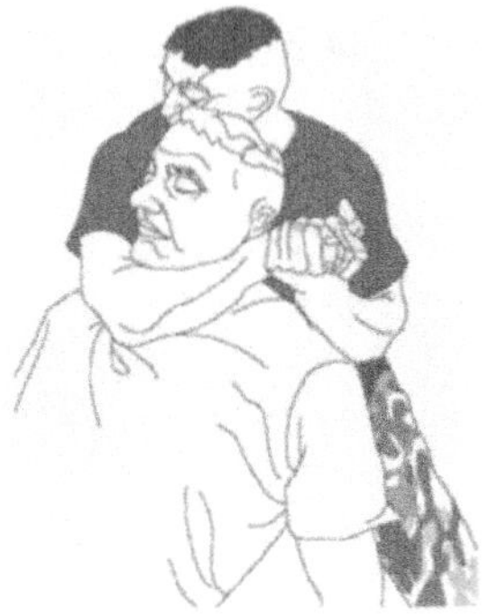

Grasp the left biceps with the right hand and place the left hand against the back of the opponent's head.

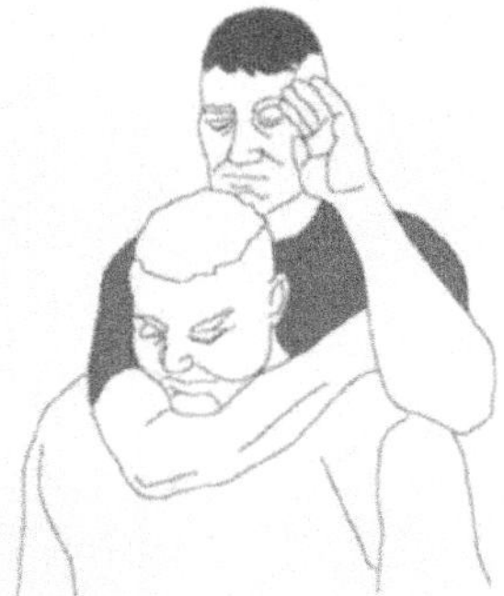

Push the opponent's head forward and down with the left hand.

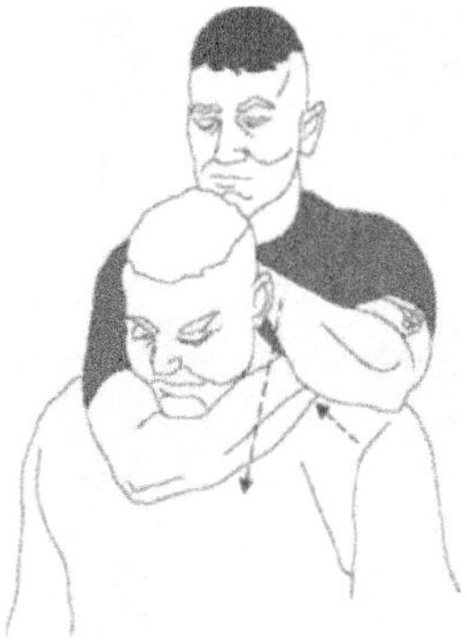

Draw the right arm in, maintaining pressure with the bicep and forearm on both sides of the opponent's neck.

3. Counters to Chokes and Holds

During a close combat situation, an opponent may apply a choke or hold on a Marine. If the opponent correctly applies a choke, a Marine quickly loses consciousness. If a choke is not executed properly, it often results in a hold, typically a bear hug or a headlock. A hold allows the opponent to control a Marine and removes the Marine's ability to attack. It is important for Marines to extract themselves from chokes and holds, regain the tactical advantage, and counter with strikes.

Although a choke causes unconsciousness in 8 to 13 seconds for a blood choke and 2 to 3 minutes for an air choke, the first movement in any counter to a choke is to clear the airway. Marines use softening techniques to loosen an opponent's grip and to clear their airway. Softening techniques are particularly effective if Marines lack the physical strength of their opponent. These techniques include groin strikes, eye gouges, foot stomps, etc. Softening techniques are not offensive; rather, they are used to loosen an opponent's hold.

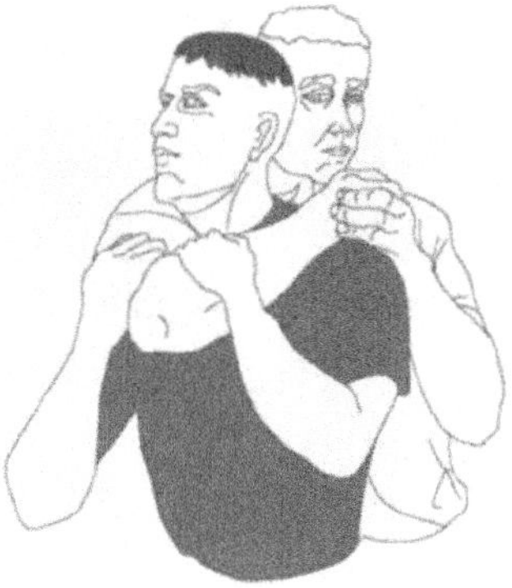

The second movement is to tuck the chin. Once the airway is clear, Marines tuck their chins to prevent the opponent from reapplying the choke.

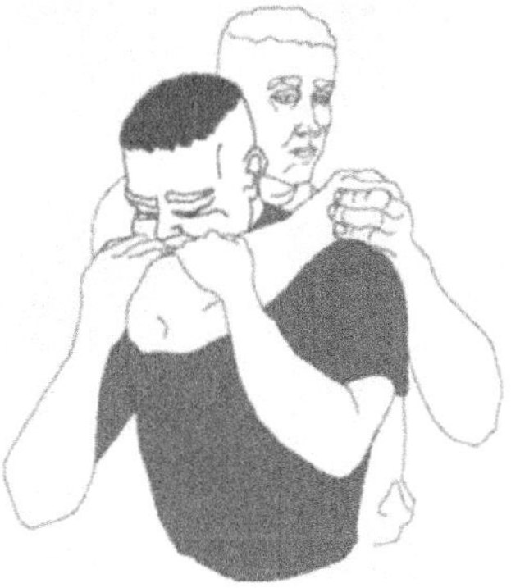

Counter to a Front Choke

Marines use a counter to a front choke when the opponent approaches from the front and uses both hands to choke a Marine around the throat. To execute the counter to the front choke, Marines—

Grasp the opponent's right forearm (where the elbow bends) with the left hand and apply downward pressure on the opponent's radial nerve with the fingers.

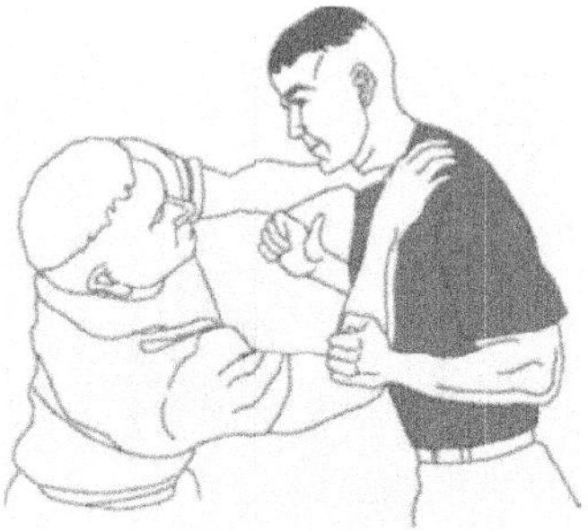

Execute a chin jab to the opponent's chin with the right hand. To generate power into the strike, bring the left foot to the outside of the opponent's right foot and rotate the hips into the strike.

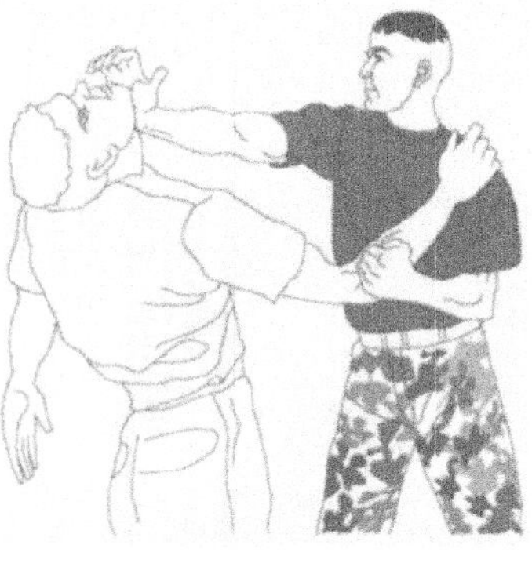

Counter to a Rear Choke

Marines execute a counter to a rear choke when the opponent approaches from the rear and puts his right arm around a Marine's throat. To execute the counter to the rear choke, Marines-

Grasp the opponent's forearm (at the radial nerve) and bicep with both hands and pull down just enough to clear the airway. Once the airway is clear, tuck the chin to protect the airway and to prevent the opponent from reapplying the choke.

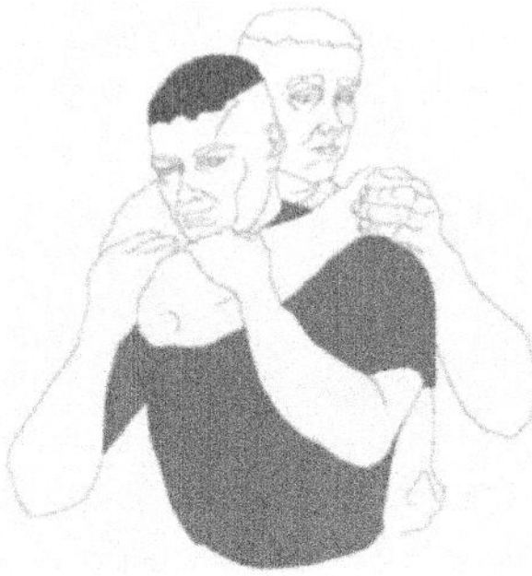

Step behind the opponent's right leg with the left foot, keeping both legs bent (almost in a squatting position).

Strike and drive the left elbow into the opponent's torso while rotating the hips and pivoting to the left, throwing the opponent back and to the ground.

Counter to a Front Headlock

Marines use a counter to a front headlock when the opponent approaches from the front and puts his right arm around the Marine's neck, bends the Marine forward, and locks the Marine's head against his hip. To execute the counter to a front headlock, Marines—

Grasp the opponent's wrist and forearm with both hands and pull down to clear the airway. Maintain control of the opponent's wrist throughout the move. Once the airway is clear, tuck the chin to protect the airway and to prevent the opponent from reapplying the choke.

Move the right hand and arm across the opponent's torso.

Step forward and to the left with the left foot at a 45-degree angle.

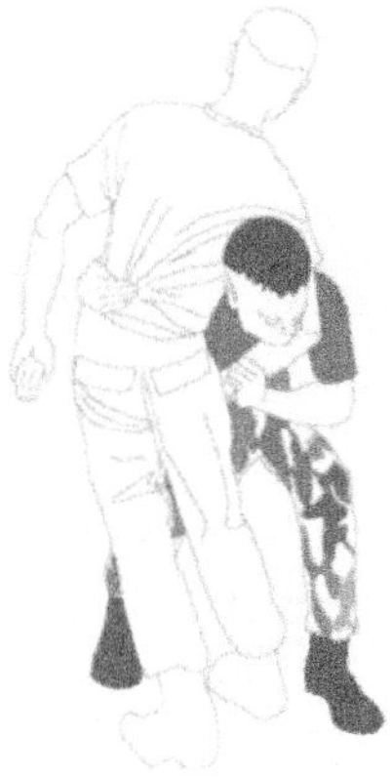

Execute a sweep with the right foot against the opponent's right leg. At the same time, push against the opponent's chest with the right arm and shoulder to generate power in the sweep.

Counter to a Rear Headlock

Marines use a counter to a rear headlock when the opponent approaches from the rear and puts his right arm around the Marine's neck, bends the Marine forward, and locks the Marine's head against his hip. To execute the counter to the rear headlock, Marines-

Grasp the opponent's wrist and forearm with the right hand and pull down to clear the airway. Once the airway is clear, tuck the chin to protect the airway and to prevent the opponent from re-applying the choke.

Reach over the opponent's right shoulder with the left arm.

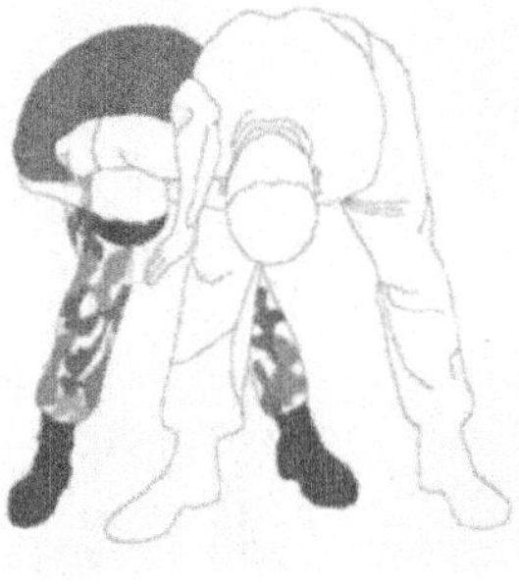

Grab any part of the opponent's face (chin, nose, eyes) and pull back while rising to a standing position.

Execute, with the right hand, a hammer fist strike to the opponent's exposed throat.

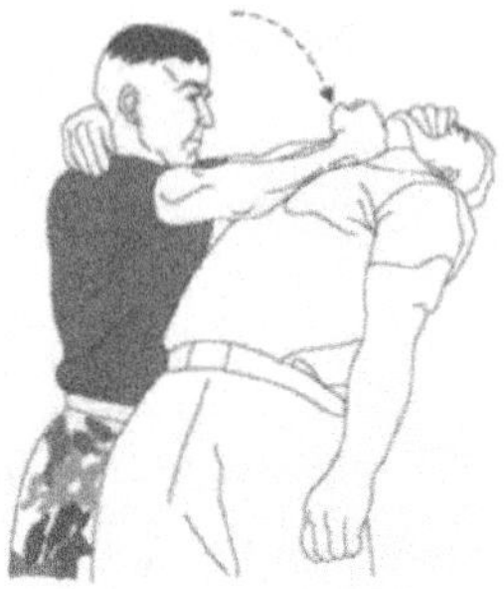

Counter to a Front Bear Hug

Marines execute a counter to a front bear hug when the opponent approaches from the front and puts both of his arms around the Marine's body, trapping the Marine's arms to the sides. To execute the counter to a front bear hug, Marines—

Step forward and to the left with the left foot at a 45-degree angle to the outside of the opponent's right leg, keeping the left leg bent.

Grasp the opponent's torso or arms to gain balance and to assist in throwing the opponent. It may be helpful to hook the opponent's right arm with the left arm.

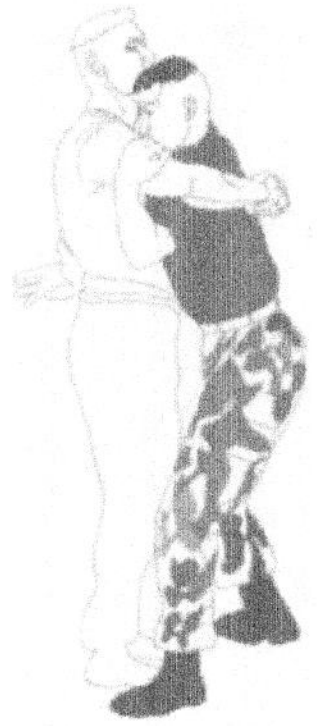

Drive the right arm and shoulder forward and, at the same time, bring the right leg forward and sweep the opponent's right leg, bringing him to the ground.

Counter to a Rear Bear Hug

Marines use a counter to a rear bear hug when the opponent approaches from the rear and puts both of his arms around the Marine's body, trapping the Marine's arms to the sides. To execute the counter to a rear bear hug, Marines—

Step behind the opponent's right leg with the left foot, keeping both legs bent (almost in a squatting position). The left side of the body should be against the opponent's.

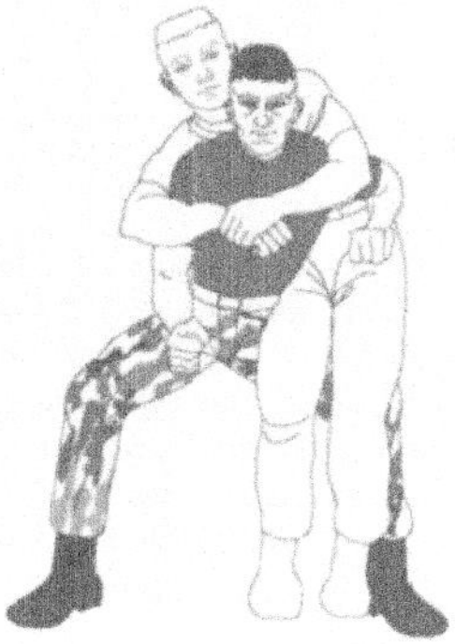

Pivot the hip, turning the body to the left and throwing the opponent back over the bent leg.

CHAPTER 7

GROUND FIGHTING

This chapter describes all techniques for a right-handed person. However, all techniques can be executed from either side.

In drawings, the Marine is depicted in woodland camouflage utilities; the opponent is depicted without camouflage. In photographs, the Marine is depicted in woodland camouflage utilities; the opponent is depicted in desert camouflage utilities.

Marines should avoid being on the ground during a close combat situation because the battlefield may be covered with debris and there is an increased risk of injury. However, many close combat situations involve fighting on the ground. The priority in a ground fight is for Marines to get back on their feet as quickly as possible. In any ground fighting scenario, Marines will usually end up in one of four positions with the opponent. The offensive positions, in which Marines have a tactical advantage, are the guard and mount. The defensive positions, which are used as counters when the opponent has the tactical advantage, are the counter to the guard and the counter to the mount. Marines can also employ chokes during ground fighting to quickly end a fight.

1. Offensive Ground Fighting

Guard Position

Marines execute a guard position when the opponent is on top and the legs are wrapped around the opponent's legs. If the opponent is on top, he may try to choke the Marine, but the Marine still has the tactical advantage because the Marine is in a position to control the situation. To execute the guard position technique, Marines—

Trap the opponent's hands on the chest by crossing the hands on the chest so the forearms are resting on the opponent's forearms. Apply pressure with the elbows or forearms.

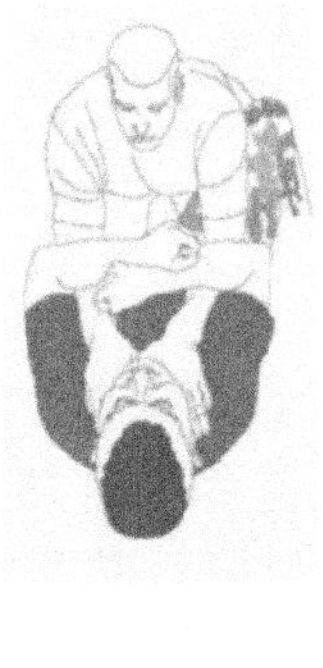

Strike the outside of the opponent's thigh with the cutting edge of the right heel. This causes the opponent to jerk to that side.

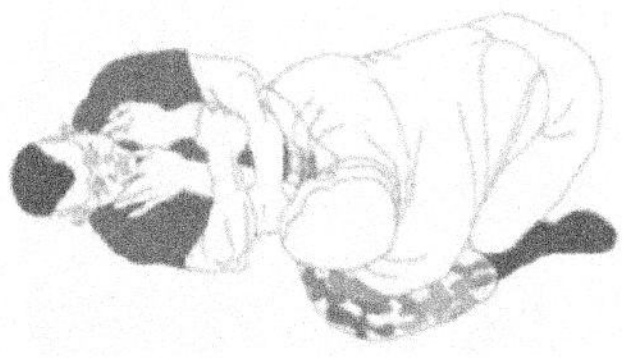

Move the head quickly to the left and swivel the hips to the right. At the same time, bring up both of the legs. Both legs are on the right side of the opponent's body.

Bring the right leg down, hooking the opponent's neck and head. Exert downward pressure to roll him over on his back. Grasp and maintain control of the opponent's left arm. Upon completion of the move, the Marine is sitting up with legs bent over the opponent while maintaining control of his arm.

Keep legs and knees bent. Maintain pressure against the opponent's neck with the back of the right foot and against his side with the left foot underneath his armpit. Squeeze the knees together, locking the opponent's arm.

Pull the opponent's arm straight up and fall back sharply, pulling his arm to the side in the direction of his little finger.

Return to standing.

Mount Position

Marines execute the mount if the opponent is lying on his back on the ground and the Marine is on top with legs wrapped around the opponent's body. This position is an offensive position because Marines are in a better position to control the opponent and to execute ground fighting techniques. To execute the mount position technique, Marines—

Grab the opponent's wrists or forearms with the hands. Hold them tightly against the torso.

Maintain control of the opponent's arms with the left arm and apply pressure to the opponent's brachial plexus (tie in) with the right hand.

Bring the left foot close to the opponent's right armpit and swing the right leg across the opponent' s head.

Keep the legs and knees bent. Maintain pressure against the opponent's neck with the back of the right foot and against his side with the left foot which is underneath his armpit. Squeeze the knees together, locking the opponent's arm.

Pull the opponent's arm straight up and fall back sharply, pulling his arm to the side in the direction of his little finger.

Return to standing.

2. Defensive Ground Fighting

Counter to the Guard

Marines use this technique if the opponent is lying on his back on the ground and the Marine is kneeling on the ground between the opponent's legs. To execute the counter to the guard position, Marines—

Strike the opponent's femoral nerve, located on the inside of the thigh, with the elbows. This forces the opponent to separate his legs.

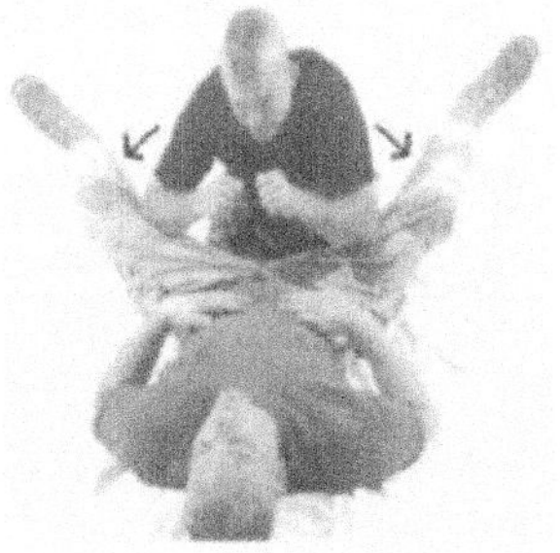

Strike the opponent's groin with the right fist.

Hook the left arm underneath the opponent's right knee from the inside. Throw the opponent's leg over the head.

Duck the head quickly and move to the left.

Return to standing.

Counter to the Mount

Marines use this technique when they are lying on their back on the ground, the opponent is mounted on top of the Marine, and the opponent's legs are wrapped around the Marine. The opponent has the tactical advantage. To execute the counter to the mount position, Marines—

Grab the opponent's gear or clothing on his upper torso and pull him down close.

Use the right arm to hook the opponent's left arm, from the inside around the outside, above his elbow. With the right foot, hook the opponent's left leg or ankle.

Draw the elbow in to bend the opponent's elbow, bringing him down close. The arm must be hooked above the opponent's elbow in order to bend it.

Strike the opponent's side with the left hand.

Push the opponent over and roll him off to the right side.

Maintain control of the opponent's hooked arm and move to a kneeling position.

Move to a standing position while maintaining pressure on the opponent's arm and using the knee to apply pressure against the opponent's elbow.

3. Ground Fighting Chokes

WARNING

During training, never execute a choke at full force or full speed. Never hold a choke for more than 5 seconds.

The priority in ground fighting is for Marines to get back on their feet as quickly as possible. Sometimes, Marines can quickly end a ground fight by executing a choke on the opponent. When performed correctly, a choke can render an opponent unconscious in as little as 8 to 13 seconds. Chokes are easily performed regardless of size or gender. The chokes performed during ground fighting are the same as those performed while standing (see chap. 6).

Ground Fighting Front Choke

The ground fighting front choke is a blood choke performed most effectively from the mount position. The front choke employs the opponent's lapels or collar to execute the choke. To execute the ground fighting front choke, Marines—

Grab the opponent's right lapel with the right hand, making certain that the knuckles or the back of the hand are against the carotid artery on the right side of the opponent's neck.

Keep the right hand pressed against the opponent's carotid artery, reach under the right arm with the left hand, and grab the opponent's left lapel, forming an X with the wrists.

Keep the right hand pressed against the opponent's carotid artery and pull the opponent's left lapel to the left with the left hand.

Ground Fighting Side Choke

The ground fighting side choke is a blood choke performed from the mount position. The ground fighting side choke is particularly effective when the opponent raises his arms and places them on the Marine's chest or throat. To execute the ground fighting side choke, Marines-

Use the left hand to parry the opponent's right arm inboard (to the inside of the opponent's reach). Bring the right arm underneath the opponent's arm and up around the front of his neck.

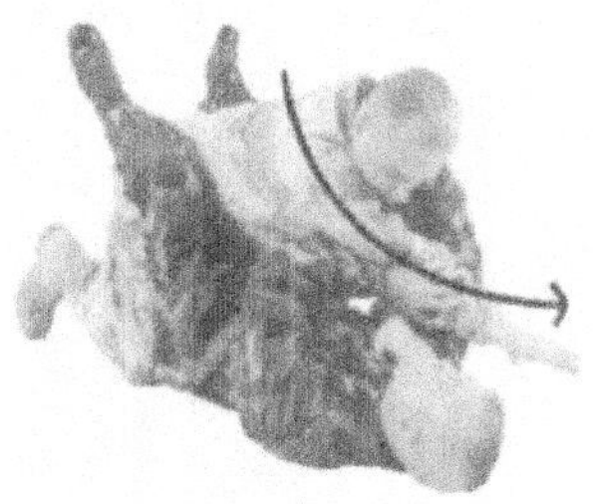

Keep the fingers extended, place the back of the forearm against the opponent's neck just below his ear and press the carotid artery on his neck. Depending on the Marine's or the opponent's body configuration, the back of the thumb or wrist may press against the opponent's neck.

Use the left hand to reach around the back of the opponent's neck and clasp the hands together.

Pull the opponent toward the chest. Use the fore-arm to exert pressure on the side of his neck. This is done by pulling the clasped hands toward the chest.

Ground Fighting Rear Choke

The ground fighting rear choke is a blood choke performed when Marines are behind the opponent. To execute the ground fighting rear choke, Marines—

Wrap the legs around the opponent with the left foot against the inside of his left thigh and the right foot against the inside of his right thigh.

Use the right arm to reach over the opponent's right shoulder and hook the bend of the arm around his neck.

Use the left hand to clasp both hands together.

Exert pressure with the biceps and forearms on both sides of the opponent's neck. This applies pressure to the opponent's carotid arteries. While maintaining pressure with the biceps and fore-arms, draw the opponent closer by drawing the right arm in. To increase the effectiveness of the choke, lean backward by arching the back and pulling the opponent back. At the same time, push the feet against the opponent's thighs.

Figure-Four Choke

A variation of the ground fighting rear choke is the figure-four choke. The figure-four choke allows Marines to gain more leverage on the rear choke. If the rear choke cannot be secured, the figure-four variation may be applied to increase the pressure of the choke on the opponent. To execute the figure-four variation of the rear choke, Marines—

Apply the ground fighting rear choke. The chest should be against the opponent's back.

Grasp the left biceps with the right hand and place the left hand against the back of the opponent's head.

Use the left hand to push the opponent's head forward and down.

Draw the right arm in while maintaining pressure with the biceps and forearm on both sides of the opponent's neck.

Increase the effectiveness of the choke by arching backward and pushing the feet against the opponent's thighs. Continue to pull the opponent back with the right arm while exerting pressure forward with the left hand against the opponent's head.

CHAPTER 8

NONLETHAL TECHNIQUES

This chapter describes all techniques for a riglit-handed person. However, all techniques can be executed from either side.

In drawings, the Marine is depicted in woodland camouflage utilities; the opponent is depicted without camouflage. In photographs, the Marine is depicted in woodland camouflage utilities; the opponent is depicted in desert camouflage utilities.

The Marine Corps' involvement in military operations other than war—e.g., humanitarian, peacekeeping, or evacuation missions—has greatly increased. These missions require skills that span the spectrum of conflict and support operations within a continuum of force. But the Corps' day-to-day existence also demands a responsible use of force. Nonlethal techniques are among the skills Marines use to apply a responsible use of force.

1. Unarmed Restraints and Manipulation

Marines operate within a continuum of force, particularly in support of peacekeeping or humanitarian types of missions. In these situations, Marines must act responsibly to handle a situation without resorting to deadly force. Unarmed restraints and manipulation techniques including joint manipulation, come-alongs, and takedowns can be used to control a subject without resorting to deadly force. Marines must train to become proficient in nonlethal techniques and to respond in a responsible manner. These techniques are referred to as compliance techniques, and they are applied in the third level in the continuum of force.

WARNING

During training, never apply the techniques for unarmed restraints and manipulation at full force or full speed. Use a slow and steady pressure to avoid injury.

Compliance Techniques

Compliance techniques are unarmed restraint and manipulation techniques used to physically force a subject or opponent to comply. Compliance can be achieved through the close combat techniques of-

Pain compliance using joint manipulation and pressure points. (Pain compliance is the initiation of pain to get compliance on the part of the subject.)
Come-along holds.

Principles of Joint Manipulation

Joint manipulation is used to initiate pain compliance and gain control of a subject. It involves the application of pressure on thejoints (elbow, wrist, shoulder, knee, ankle, and fingers). Pressure is applied in two ways:

In the direction in which the joint will not bend. For example, joints such as the knees and elbows only bend in one direction and when pressure is applied in the opposite direction, pain compliance can be achieved. Beyond the point where the joint stops naturally in its range of movement (i.e., it no longer bends).

Since each joint has a breaking point, Marines should apply slow steady pressure only until pain compliance is reached. Continued pressure will break the joint and may escalate the violence of the situation.

Joint manipulation also uses the principle of off-balancing. A subject can be better controlled when he is knocked off balance.

Wristlocks

A wristlock is a joint manipulation that can be applied in a number of ways to achieve pain compliance. The wrist rotates in a number of directions and will bend in a single direction until its movement stops naturally. In a wristlock, pressure is exerted beyond that point by bending or twisting the joint. A wristlock is executed when an opponent tries to grab Marines or is successful in grabbing Marines or their equipment. A wristlock can also be performed by Marines if they wish to initiate control of an opponent.

Basic Wristlock. A basic wristlock is executed when Marines grab the opponent's left hand with the right hand. To execute the basic wristlock, Marines—

Use the right hand to grab the opponent's left hand by placing the thumb on the back of the opponent's hand so that the Marine's knuckles are facing to the left.

Hook the fingers across the fleshy part of the opponent's palm below the thumb. The fingers are used to anchor the hand so leverage can be applied to twist and bend the joint.

Exert downward pressure with the thumb to bend the opponent's joint. Rotate the opponent's hand to the right to twist the joint.

Use the left hand to further control the opponent.

Step in to the opponent to keep the opponent's hand in close to the body to control him and provide more leverage on the wristlock.

Note: When executing the basic wristlock with the left hand, the Marine grabs the opponent's hand so that the Marine's knuckles are facing to the right, and then rotates and twists the opponent's hand to the left.

Reverse Wristlock. A reverse wristlock is executed when Marines grab the opponent's right hand with the right hand. To execute the reverse wristlock, Marines—

Place the right palm on the back of the opponent's right hand and wrap the fingers across the fleshy part of his palm below his little finger.

Twist the opponent's hand to the right while stepping in to place his hand against the chest. Apply downward pressure on the opponent's hand against the chest. Leave the opponent's hand on the chest to fully control the subject and to gain leverage.

Lean forward to use body weight to add additional pressure to the joint.

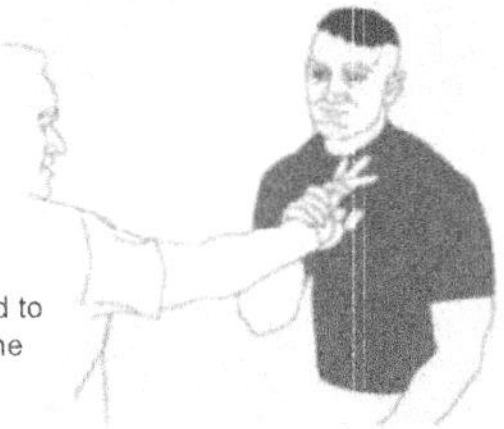

Use the left hand to further control the opponent.

Two-Handed Wristlock. Both hands can be used in the wristlock to maximize the leverage and pressure needed to bend and twist the joint. To execute the two-handed wristlock, Marines—

Place both thumbs on the back of the opponent's hand, thumbs crossed.

Hook the fingers of both hands around the fleshy part of the opponent's palm on both sides of his hand.

Step into the opponent and apply pressure downward on the back of his hand to bend the joint and rotate his wrist away from the body to twist the joint.

Enhanced Pain Compliance on Wristlock. Enhanced pain compliance techniques are applied in the third and fourth levels in the continuum of force. Additional pain be applied to a wristlock by-

Adding downward pressure to the elbow with the other hand or elbow by using the fingers to pull in on the opponent's radial nerve located on the inside of the forearm. When pressure is added to the opponent's radial nerve, his direction can be controlled.

Applying pressure against the opponent's finger joint to bend it in a direction it cannot bend (i.e., splitting the fingers).

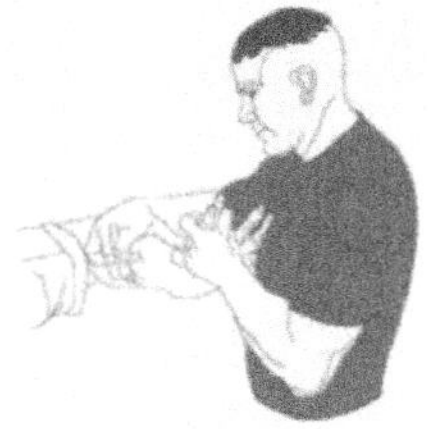

Come-Along Holds

Marines use a come-along hold to control and move an opponent.

Escort Position. A common come-along hold is the escort position. To execute the escort position, Marines—

Face the opponent. Use the left foot to step forward at a 45-degree angle. Turn to face the right side of the opponent.

Use the right hand to firmly grasp the opponent's right wrist. With the left hand, firmly grasp the opponent's right triceps.

Position the opponent's controlled arm diagonally across the torso, keeping his wrist against the right hip. The Marine should be standing to the right of and behind the opponent.

Note: This technique works well when escorting an opponent on either the right or left side. Take caution when escorting an opponent by ensuring his controlled hand is not in a position to grab the holstered weapon. The preferred escort position is from the Marine's left side, so that the opponent is kept further away from the weapon.

Wristlock Come-Along. To execute the wristlock come-along, Marines-

Use the left hand to execute a basic wristlock. Incorporate the right hand in a two-handed wristlock for more control.

Maintain pressure on the opponent's wrist with the right hand, step forward, and pivot around to stand next to the opponent.

Release the left hand, quickly reach under the opponent's arm from behind, and grab his hand.

Use the left hand and apply downward pressure on the opponent's wrist.

Controlling Technique. The following controlling technique is used when an opponent grabs the Marine's wrist. To execute the technique, Marines---

Trap the opponent's hand with the palm of the other hand.

Rotate the opponent's trapped hand up and on his forearm while maintaining downward pressure on his trapped hand.

Apply downward pressure with both hands until the opponent is taken to the ground.

Armbars

An armbar is a joint manipulation in which pressure is applied on a locked elbow, just above the joint, in the direction the joint will not bend. An armbar has to be locked in quickly, but still requires a slow, steady pressure to gain compliance.

Basic Armbar. To execute a basic armbar, Marines-

Use the right hand to grab the opponent's right wrist.

Bring the left hand down on or above the opponent's elbow joint. To gain additional leverage, pivot to face the opponent.

Use the left hand to apply downward pressure on or above the opponent's elbow joint while pulling up on his wrist.

Armbar from a Wristlock. To execute an armbar from a wristlock, Marines-

Use the right hand to grab the opponent's right hand and execute a reverse wristlock.

Bring the left hand down on or above the opponent's elbow joint.

Use the left hand to apply downward pressure on or above the opponent's elbow joint while pulling up on his wrist.

Takedowns

A takedown is used to bring an opponent to the ground to further control him.

Takedown From a Wristlock Come-Along. To take the opponent to the ground from a wristlock come-along, Marines—

Use the right foot to push down on the opponent's calf or Achilles tendon.

Maintain control of the opponent's wrist and elbow and apply a slow, steady pressure to bring him to the ground.

Armbar to a Takedown. This technique is used to take a noncompliant opponent to the ground from an armbar. To execute the armbar takedown, Marines—

Use the right hand to execute a reverse wristlock.

Bring the left hand or forearm down on or above the opponent's elbow joint.

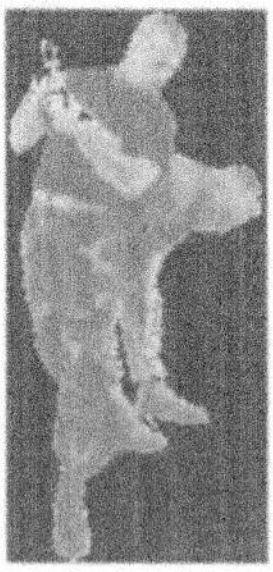

Pivot so the back is facing the opponent and, at the same time, lift the left elbow and slide the body so it is against the opponent, placing the armpit high above the opponent's elbow joint.

Lean back, placing the body weight on the opponent's arm until he complies or is taken to the ground.

Note: This technique may break the opponent's arm. Therefore, this technique should not be employed if the objective is a nonlethal takedown.

Wristlock Takedown. This technique is used to take a noncompliant opponent to the ground from a basic wristlock and to put him in a position where he can be handcuffed, if necessary. To execute the takedown, Marines-

Use the right hand to execute a basic wristlock.

Incorporate the left hand in a two-handed wristlock.

Apply downward pressure on the wristlock, pivot on the ball of the right foot, and quickly turn to the right to take the opponent to the ground.

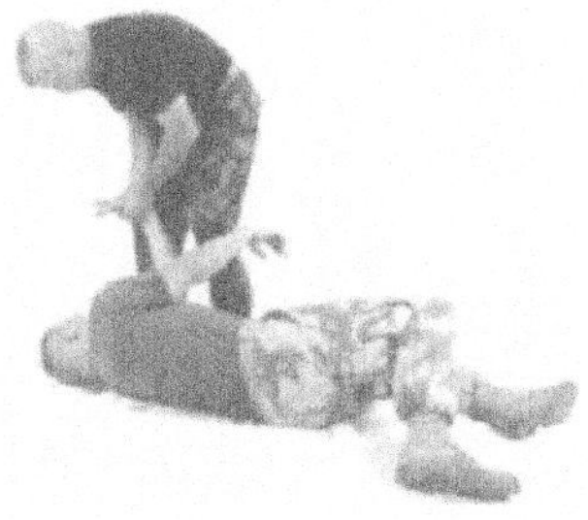

Continue to apply pressure on the wristjoint as the opponent lands on his back with his arm straight in the air.

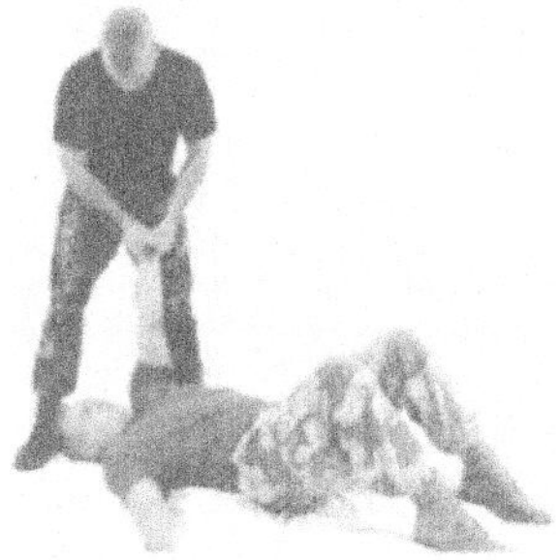

Slide the left foot under the opponent's back, underneath his armpit.

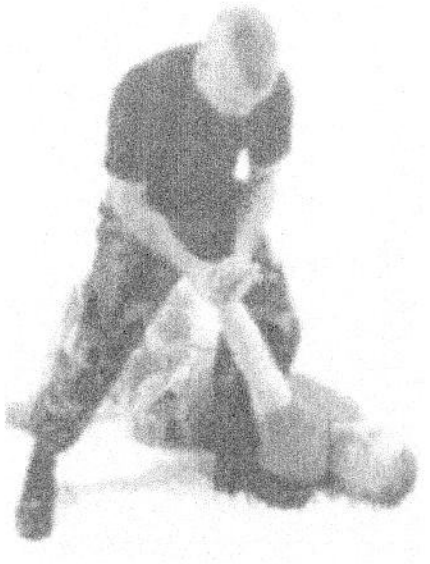

Apply pressure with the knee against the opponent's triceps while pulling back on his arm.

Continue applying pressure with the knee on the opponent's arm. Pivot and step around the opponent's arm to roll him on his stomach.

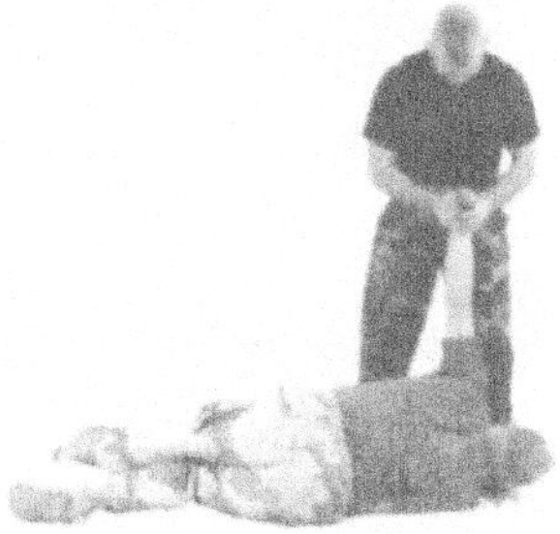

Kneel down with one knee on the opponent's back. The other knee is placed on the opponent's neck and shoulder on either side of his arm. Apply inward pressure with the knees to lock his arm in place.

Tell the opponent to put his other hand in the middle of his back. Bring the opponent's controlled hand to the center of his back.

Escort Position Takedown. This technique can be used to control a noncompliant opponent from the escort position. To execute the escort position takedown, Marines—

Lock the opponent's arm straight across the body while rotating his wrist away from the body.

Use the left hand or forearm and apply downward pressure above the opponent's elbow where the triceps meet.

Step back with the right foot and, keeping the opponent's hand controlled against the hip, pivot to the right while continuing to apply downward pressure on his arm to bring him to the ground.

Note: This technique works well when escorting a subject on the Marine's right or left side. When taking down a subject from the right side, step back and pivot to the left.

2. Nonlethal Baton

A baton or nightstick can be an effective compliance tool when used correctly. Batons can be used defensively (blocking), offensively (striking), and as a restraining device when needed. In the fourth level of the continuum of force (assaultive [bodily harm]), defensive tactics include baton or nightstick blocks and blows. Blows to the head or other bony parts of the body are considered deadly force. When deadly force is not authorized, Marines must be able to employ blocks, strikes, and restraints with the baton with the minimum of force.

Grips

One-Handed Grip. To execute the one-handed grip, Marines—

Use the right hand to grasp the lower end of the baton, about 2 inches from the end.

Wrap the thumb and index finger around the baton so they are touching one another. The grip on the baton should be firm, but natural.

Two-Handed Grip. To execute the two-handed grip, Marines-

Use the right hand to grasp the lower end of the baton, about 2 inches from the end. Wrap the thumb and index finger around the baton so they touch one another.

Use the left hand to grasp the upper end of the baton, palm down, about 2 inches from the end. The hands should be approximately 10 to 12 inches apart.

Stance and Method of Carry

The basic warrior stance serves as the foundation for initiating nonlethal baton techniques. The method of cany provides effective defensive positions with a wide range of options to control a combative opponent.

One-Handed Carry. To execute the one-handed carry, Marines—

Grip the baton using the one-handed grip.

Elevate the baton, with the gripping hand at a level between the belt and shoulder.

Keep the left hand in the position of the basic warrior stance.

Two-Handed Carry. This carry is effective for blocks. To execute the two-handed carry, Marines-

Grip the baton using the two-handed grip.

Elevate the baton, with the left hand higher than the right hand.

Orient the weapon toward the subject.

Movement

In a nonlethal confrontation, movement may be made to create distance between Marines and the opponent or to close the gap to control him. When facing an opponent, Marines move in a 45-degree angle to either side of the opponent. Moving at a 45-degree angle is the best way to both avoid an opponent's strike and to put Marines in the best position to control the opponent.

Blocking Techniques

One-handed blocks are used when carrying the baton in the one-handed carry. The same one-handed blocks used in combative stick techniques (see page 3-6) apply in nonlethal baton engagements. The blocks include blocks for a vertical strike, a forward strike, and a reverse strike. Because the baton is often carried with two hands, there are also two-handed blocks that are effectively used from this carry. Two-handed blocks are discussed in the following subparagraphs.

High Block. Marines execute a high block to deter a downward vertical attack directed at the head and shoulders. To execute the high block, Marines—

Raise the baton up to a level even with or above the head. The baton should be in a horizontal position to block the blow. The fingers of the left hand should be open and behind the baton.

Place the baton perpendicular to the opponent's striking surface to absorb the impact of the blow.

Bend the elbows slightly to help absorb the impact of the blow. The arms should give with the strike of the blow.

Low Block. Marines execute a low block to deter an upward vertical attack directed at the abdomen, groin, or torso. The opponent's blow can be delivered by a foot, knee, or fist. To execute the low block, Marines—

Lower the baton to a level even with or below the groin. The baton should be in a horizontal position to block the blow. The fingers of the left hand should be open and behind the baton.

Place the baton perpendicular to the opponent's striking surface to absorb the impact of the blow.

Bend the elbows slightly to help absorb the impact of the blow. The arms should give with the strike of the blow.

Right Block. Marines execute a right block to deter an attack directed at the head, neck, flank, or hip. The opponent's blow can be delivered by a foot, knee, fist, or elbow. To execute the right block, Marines-

Thrust the baton in a vertical position to the right side.

Pivot to the right by stepping with the left foot and pivoting off the ball of the right foot. Rotate the hips and shoulders into the direction of the block. The fingers of the left hand should be open and behind the baton.

Place the baton perpendicular to the opponent's striking surface to absorb the impact of the blow.

Bend the elbows slightly to help absorb the impact of the blow. The arms should give with the strike of the blow.

Left Block. Marines execute a left block to deter an attack directed at the head, neck, flank, or hip. The opponent's blow can be delivered by a foot, knee, fist, or elbow. To execute the left block, Marines—

Thrust the baton in a vertical position to the left side.

Pivot to the left by stepping with the right foot and pivoting off the ball of the left foot. Rotate the hips and shoulders into the direction of the block. The fingers of the left hand should be open and behind the baton.

Place the baton perpendicular to the opponent's striking surface to absorb the impact of the blow.

Bend the elbows slightly to help absorb the impact of the blow. The arms should give with the strike of the blow.

Middle Block. Marines execute either a left or right block to deter an attack directed at the face, throat, chest, or abdomen. To execute the middle block, Marines—

Thrust the baton in a vertical position straight out in front of the body. The fingers of the left hand should be open and behind the baton.

Place the baton perpendicular to the opponent's striking surface to absorb the impact of the blow. The baton should be held with the left hand forward of the right.

Bend the elbows slightly to help absorb the impact of the blow. The arms should give with the strike of the blow.

Restraining Technique

The strong-side armlock is used to restrain an opponent who is not compliant. To execute the strong-side armlock, Marines—

Use the right hand to run the baton up under the opponent's left armpit, parallel to the ground.

Use the right foot to step forward at a 45-degree angle to the left side of the opponent. The baton is across his forearm.

Use the right hand to drive the baton forward and up so the action bends the opponent's arm behind his back. At the same time, continue moving around him to get behind him.

Place the baton on the opponent's forearm with the thumb and/or fingers and apply pressure to his forearm. At the same time, grasp the other end of the baton with the left hand.

Pull up on the low end of the baton with the right hand. At the same time, push down on the top end of the baton with the left forearm, reaching around with the left hand to grasp the opponent's biceps or shoulder.

Continue exerting downward pressure with the left forearm while pulling back on the opponent's biceps with the left hand. This places the opponent in a position where he is controlled and can be moved.

Apply pressure with the foot against the bend in the opponent's leg above his calf. This lowers the opponent to the ground rather than throwing him to the ground and risking severe injury.

Striking Target Areas

Marines must avoid striking an opponent in the head, neck, or other bony parts with the baton because this is considered deadly force and can lead to serious bodily injury or death. Instead, the legs, arms, and buttocks are target areas that are considered nonlethal.

Legs. Primary targets are the thighs and lower legs. Avoid striking the knee and ankle joints because this can cause permanent damage.

Arms. Primary targets are the upper arms. Avoid striking the shoulder and elbow and wrist joints because this can cause permanent damage.

Buttocks. Primary targets are the buttocks. Avoid striking any other part of the torso, including the chest, rib cage, spine, tail bone, and groin because strikes to these areas can cause permanent damage or death.

One-Handed Striking Techniques

One-Handed Forward Strike. A forward strike follows either a horizontal line or a downward diagonal line using a forehand stroke. To execute the one-handed forward strike, Marines-

Stand facing the opponent with the baton carried in a one-handed cany.

Place the right hand palm up, swing the baton from right to left, and make contact with the opponent.

One-Handed Reverse Strike. A reverse strike follows either a horizontal line or a downward diagonal line using a backhand stroke. To execute the one-handed reverse strike, Marines—

Stand facing the opponent with the baton carried in a one-handed carry.

Bend the right arm and cross the arm to the left side of the body. The baton should be close to or over the left shoulder.

Place the right hand palm down, swing the baton from left to right, and make contact with the opponent.

Two-Handed Striking Techniques

Two-Handed Forward Strike. This strike is an effective follow-up to a middle block or left block. To execute the two-handed forward strike, Marines—

Pull back with the left hand while driving the right hand forward toward the opponent. The baton should be horizontal to the ground. Power is generated by stepping forward with the right foot and rotating the right hip and shoulder into the strike.

Contact the opponent with the end of the baton.

Two-Handed Reverse Strike. This strike is an effective follow-up to a middle block or left block. To execute the two-handed reverse strike, Marines—

Pull back with the right hand while driving the left hand forward toward the opponent. The baton should be horizontal to the ground. Power is generated by stepping forward slightly with the left foot and rotating the left hip and shoulder into the strike.

Contact the subject with the end of the baton.

Front Jab. This strike is effective for countering a frontal attack. It can also be executed as a quick poke to keep a subject away. To execute the front jab, Marines thrust both hands forward in a quick jab. The baton is held either horizontal to the ground or at a slight downward angle.

Rear Jab. This strike is effective for countering a bear hug from the rear. To execute the rear jab, Marines thrust both hands rearward in a quick jab. The baton is held either horizontal to the ground or at a slight downward angle.

APPENDIX A

PUGIL STICK TRAINING

A pugil stick is a training device used to simulate a rifle bayonet so that effective, but safe, training can be conducted to build proficiency of rifle bayonet techniques. Pugil stick training builds on the techniques used to throw punches. Pugil stick training is the only "full contact" training provided to Marines in the Close Combat Program. Pugil stick training teaches Marines to function when faced with stress and violence, and it prepares them to deliver a blow and take a blow. It also provides them with the physical and mental skills vital to success on the battlefield.

1. Pugil Stick Training

Design

A pugil stick consists of a stick wrapped in padding at both ends that can be gripped like a rifle. The pugil stick is approximately the same weight and length of an unloaded rifle with a bayonet attached.

Holding the Pugil Stick

The pugil stick is held in the same manner as the sen ice rifle. All movements come from the basic warrior stance. To hold the pugil stick correctly, Marines—

Use the right hand to grasp the lower end of the pugil stick overhanded.
Use the left hand to grasp the upper end of the pugil stick underhanded.
Use the right forearm to lock the lower end of the pugil stick against the hip.
Orient the blade end of the pugil stick toward the opponent.

Safety Equipment

The following safety equipment must be worn during any pugil stick bout.

Groin Protection. Groin protection protects the groin from an accidental blow. It should be pulled high around the waist to protect the groin area, with the concave portion against the body.

Flak Jacket. The flak jacket provides protection to the body. It is worn completely fastened.

Neck Roll. The neck roll prevents whiplash if Marines receive a blow to the head. The neck roll further supports the head and protects the neck from blows. The neck roll is worn above the flak jacket and below the helmet. The tied end of the neck roll faces the front.

Helmet. A regulation football helmet protects the face and head. The helmet must fit snugly and the chin straps must be adjusted and snapped.

Mouthpiece. The mouthpiece is worn on the upper teeth to protect the teeth.

Gloves. Marines wear gloves to protect the hands if pugil sticks do not have gloves built-in.

Pugil Stick Screening

Prior to pugil stick training, instructors must ask participating Marines the following questions. If a Marine answers "yes" to any question, he must see the corpsman or a competent medical authority who evaluates him and determines whether or not the he can participate in the training.

Are you on light duty?
Arc you restricted to running shoes by a corpsman or a doctor?
Have you fought in a pugil stick bout within the last 7 days?
Have you received a blow to the head within the last 7 days?
Have you had a concussion within the last 6 months?
Have you had dental surgery within the last 24 hours?
Do you have stitches or staples on your body?
Have you had a shoulder or head injury within the last 5 years?
Are you taking a prescription drug?
Do you have an ear infection or current sinus infection?
Have you had a broken bone within the last 6 months?

Safety Personnel

The following safety personnel are required to conduct pugil stick training:

One close combat instructor must officiate the bout.
One close combat instructor trainer, commissioned officer, or staff noncommissioned officer must be in the training area to serve as range safety officer.
One corpsman must be in the training area.

Training Area

To prevent injury, Marines train on areas with soft footing (i.e., sand or grass). Training mats are not recommended because feet can stick to the mats, prohibiting movement or causing joint injuries by twisting a knee. Bouts should not take place on a hard surface area; e.g., a flight deck or parking lot. A boxing ring may be used to conduct pugil stick bouts; ring dimensions can vary as long as there is ample room to execute the techniques in the training area.

Second Impact Syndrome

Second Impact Syndrome occurs when a second blow to the head produces a second concussion that occurs within 1 week following a previous concussion (before recovery from the first concussion). Second Impact Syndrome causes rapid brain swelling and can cause death. Therefore, there must be 7 days between pugil stick bouts to reduce the risk of severe injury resulting from Second Impact Syndrome. The 7 day separation between pugil stick bouts significantly reduces the possibility of injury, particularly in someone who may have suffered a brain injury or concussion but shows no symptoms.

Any Marine who experiences headaches or the following symptoms after training must be examined by appropriate medical personnel:

Blurred vision.
Ringing in the ears.
Dilation of the pupils.
Slurred speech.
Bleeding from ears or mouth.
Swelling in head or neck area.
Any unnatural discoloration in head or neck area.

The Marine should not be allowed to participate in pugil stick training or any other activity where a heavy blow might be sustained for a minimum of 7 days after the headache or other symptoms have subsided.

Safety Measures

The following safety measures must be followed:

Gear must be worn properly throughout training.
Contact lenses or glasses will not be worn.
False teeth will be removed from the mouth.
Nothing will be worn around the neck except the neck roll.

Competition among groups of Marines is authorized as long as it does not overshadow training objectives or compromise safety procedures.

Safety and proper techniques are paramount.

Safety is more important than competition.

2. General Rules and Regulations Governing Pugil Stick Bouts

Instructors and Support Personnel Requirements

At a minimum, one close combat instructor observes a pugil stick bout. For safety purposes, it is better to have two instructors judging a bout because each instructor can fully observe each of the fighters. The best position for observation is to the right of a fighter. This allows the instructor to see the fighter's facial expression and body movement. The instructor's position must not interfere with the fight.

In addition, one commissioned or staff noncommissioned officer (to serve as range safety officer) and one corpsman will be in the training area.

Prior to the Bout

Prior to the bout, Marines—

Are paired according to height, weight, and gender.

Wear the proper safety equipment.

Once Marines have properly donned the safety equipment, they wait for instructions from the close combat instructor. On the command of the close combat instructor, a pair of Marines will enter the ring. Once in the ring, the close combat instructor inspects each Marine for the proper safety equipment.

Note: There should be enough gear so that when two Marines are training, two other Marines can be donning safety equipment for the next bout.

During the Bout

The bout begins when the close combat instructor blows the whistle. All strikes are directed above the waist. Upon hearing a whistle blast, all fighting immediately ceases.

Stopping a Bout

There are two reasons for stopping a bout: delivery of a scoring blow or an unsafe condition. The close combat instructor trainer, close combat instructor, or the range safety officer may stop the bout at any time an unsafe condition is observed.

Scoring Blow. A scoring blow is an offensive technique delivered to a vulnerable area of an opponent with sufficient force and precision to be considered as a disabling or killing blow. Scoring blows are not judged solely on the degree of force with which a blow is delivered, but on the accuracy and techniques employed. A scoring blow is defined as—

A straight thrust with the blade end of the weapon (red end of the pugil stick) to the opponent's face mask or throat.

A slash to the side of the opponent's helmet (below the ear) or neck with the red end of the pugil stick.

A heavy blow to the opponent's head with an authorized technique (i.e., buttstroke, smash) using the butt (black end) of the pugil stick.

When a scoring blow is delivered, the close combat instructor blows the whistle to stop the bout.

Unsafe Condition. The bout will be stopped as soon as an unsafe condition exists. An unsafe condition exists when a Marine is unable to defend himself, loses his balance and falls down on one or two knees or falls down completely, shows instability (e.g., buckling at the knees), loses muscular tension in his neck and his head snaps back or to one side, or appears disoriented. An unsafe condition also exists if a Marine lets go of one end of the pugil stick, the equipment (e.g., helmet, neck roll) falls off, or a Marine fails to use the proper techniques.

If any of these conditions occur—

The close combat instructor trainer, close combat instructor, range safety officer, or anyone supervising training stops the bout and separates the two Marines.
The corpsman evaluates the possibly injured Marine to determine if the fight can continue. The corpsman—

- Ensures that he is alert and responsive, both verbally and physically.
- Talks to him to see if he is coherent.
- Makes certain he comprehends and replies to verbal questions.
- Makes certain his speech is not slurred.
- Checks his physical signs.
- Ensures his eyes are focused and not dazed or glazed.
- Ensures his legs are not wobbly or shaky.

The pugil stick is not used as a baseball bat. The use of unauthorized techniques will result in expulsion from the bout. Only techniques taught in bayonet training are authorized. These techniques include-

Straight thrust.
Buttstroke (horizontal and vertical).
Smash.
Slash.
Parry.
Blocks (low, high, left, right).

After the Bout

A third whistle blast is used prior to identifying the Marine who delivered the killing blow.

A fourth whistle blast is used prior to demonstrating the killing technique used by the Marine who won the bout.

3. Directions for Making Pugil Sticks

Close combat instructor trainers are taught to make pugil sticks to standard specifications at school. It is recommended that all pugil sticks be made under the supervision of a close combat instructor trainer. Until the Marine Corps acquires a standardized pugil stick that units can order through the supply system, instructors use the following information and instructions to make pugil sticks locally.

Supplies

The following supplies are needed to make one pugil stick:

One 45" by 2" circular, oak, wood dowel.
Four 9" wide by 27" long foam pad pieces to make two inner handguards of double thickness.
Four 11" wide by 19" long foam pad pieces to make two outer handguards of double thickness.
Canvas material to cover inner and outer handguards.
One foam pad cut 1/4" thick by 7" wide by 13" long to make the centerguard.
One 8" wide by 14"long piece of canvas material to make a pugil stick sock.
Canvas material to make two pugil stick end caps. Ideally, two different color fabrics should be used to simulate the bayonet and buttstock ends of a rifle. Common colors for end caps are black and red. Each end cap is a different color to help the officiating certified close combat instructor determine the scoring blow during the pugil stick bout.
Two 1/4" thick by 11" wide by 54" long foam pad pieces to make two ends.
Two foam pads cut 1/4" thick by 3" wide by 14 1/2" long to make two end plug inserts.
One can of 3M Photo MountQ spray adhesive (NSN 8040-01-120-4009)to adhere foam padding to the stick. (Spray adhesive is available from self-service.)
One roll of duct tape or riggers tape (NSN 7510-00-266-5016) to protect foam pad and

canvas from wear and to reinforce gluing action of spray adhesive. (Duct tape or riggers tape is available from self-service.)
One roll para-cord or 5-50 cord to tie both the inner and outer handguards together.
One roll of thread strong enough to sew canvas material together.
Sixteen grommets, one for each of the four corners of the inner and outer handguards.
One foam sleeping mat (NSN 8465-01-109-3369). Use unserviceable foam sleeping mats obtained from Defense Rcutilization and Marketing Office.

Tools

The following tools are needed to make a pugil stick:

Tape measure, ruler, or yardstick.
Scissors.
Grommet machine.
Knife.
Sewing machine.

Directions

To make a pugil stick, Marines perform the following steps.

Preparing Materials

Step 1. Cut foam sleeping mat into the dimensions listed above. Cut mats to double thickness.

Step 2. Use spray adhesive to adhere the two 9" by 27" pieces of foam together. Use the spray adhesive to adhere the two 11" by 19" pieces of foam together. Let adhesive dry before using the double-thick foam for handguards.

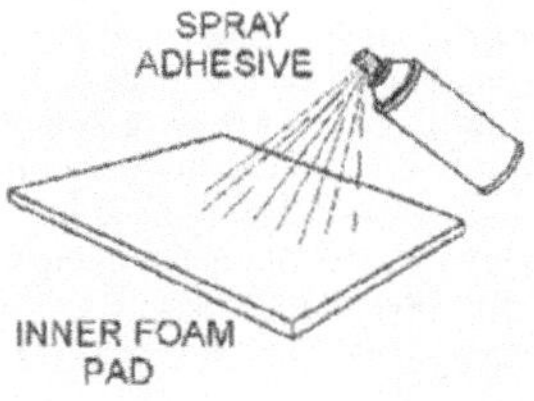

Step 3. Cut canvas material to make two coverings for the inner handguards. Cut material to make two coverings for the outer handguards. Two inner handguards and two outer handguards are required for each pugil stick.

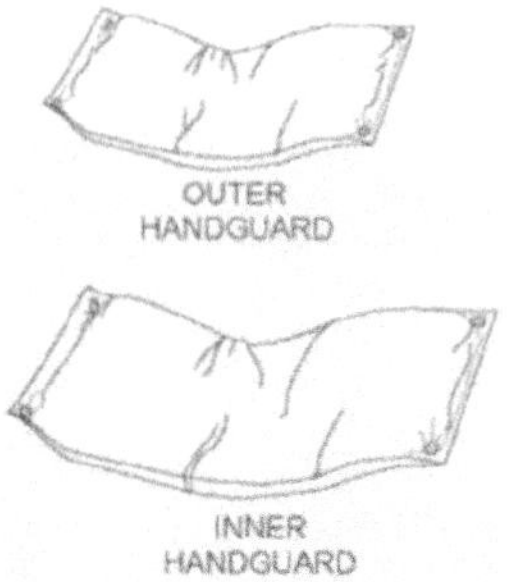

Step 4. Use a sewing machine to make a sleeve out of the 9" by 27" canvas. Sew down three sides. Insert foam pad inside the canvas sleeve and then sew down the fourth side. Use the grommet machine to create holes reinforced with metal grommets in each of the four corners. Once finished, the inner handguards arc complete.

Note: It is recommended that the Fabric Repair Shop be used to sew canvas sleeves and apply grommets.

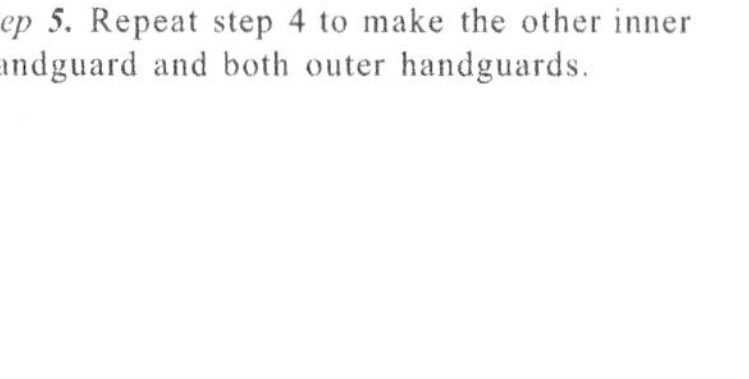

Step 5. Repeat step 4 to make the other inner handguard and both outer handguards.

Constructing Ends, End Plugs, and End Caps

Step 1. Apply spray adhesive on approximately 8 inches of one end of the dowel. This adhesive serves to glue the foam pad to the wooden dowel.

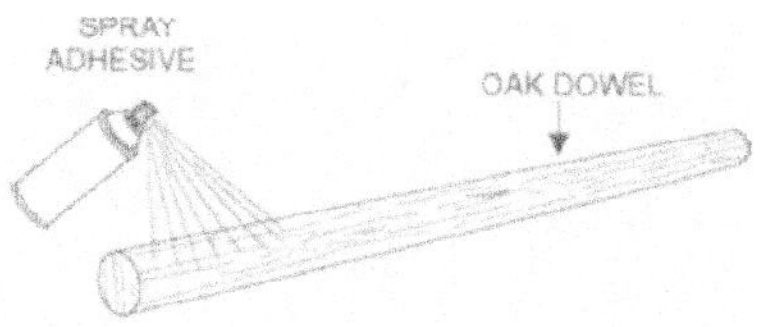

Step 2. Apply spray adhesive on one side of the 11" by 54" foam pad. This causes the foam pad to stick to itself when it is rolled around the end of the wooden dowel. Ensure 3 inches of the rolled foam pad extends past the wooden dowel for stability.

Step 3. Apply spray adhesive on one side of 3" by 14 1/2" foam pad. Roll pad into a tight cylinder.

Step 4. Apply spray adhesive on the exposed end of the dowel and fill the hole in the end with the end plug. The end plug prevents the wooden dowel from protruding beyond the edge of the rolled foam padding. This is an important safety precaution to prevent injury.

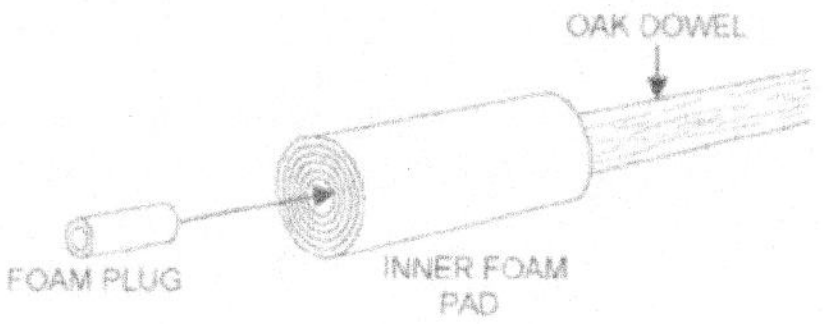

Step 5. Use duct tape or riggers tape to secure and reinforce the foam pad once the end plug is inserted in the end.

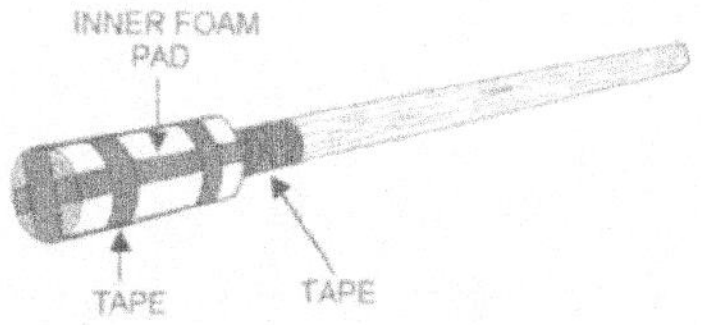

Step 6. Measure the dimensions of the ends (typically, the length of the cylinder should be 12 inches long to allow for taping). Cut and sew canvas material in the shape of a cylinder with one open end. Once complete, this becomes the end cap. The end cap is a colored sleeve that covers the rolled and taped foam pad at each end of the pugil stick.

Note: It is recommended that the Fabric Repair Shop be used to make the end cap.

Step 7. Slide one of the end caps over the rolled foam pad and tape it to the dowel.

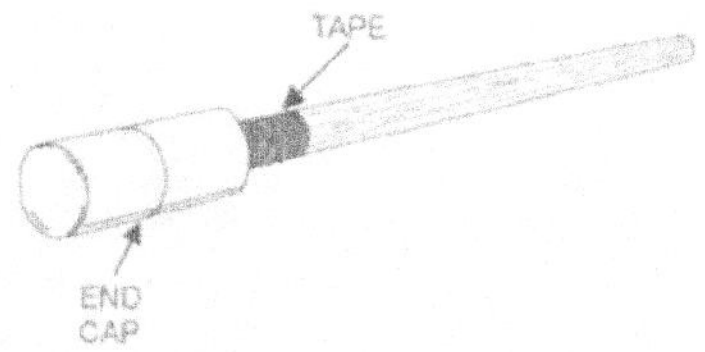

Building the Centerguard

Step 1. Spray one side of the 7" by 13" foam pad with spray adhesive and wrap it tightly around the center of the dowel.

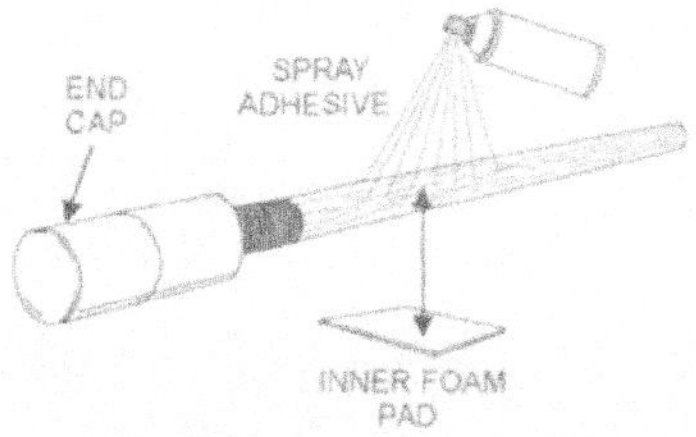

Step 2. Wrap duct tape or riggers tape around the foam pad to secure.

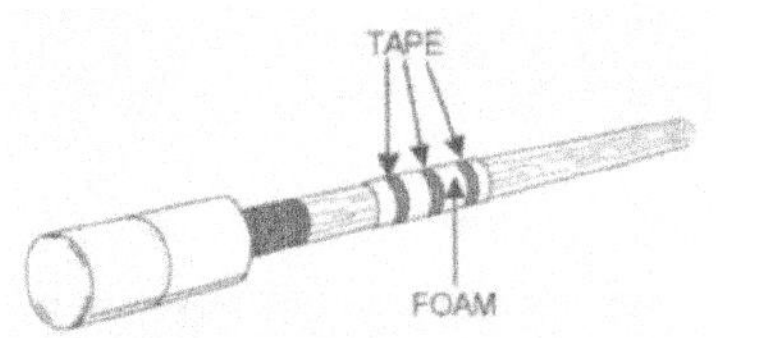

Step 3. Use a sewing machine to make a tubular pugil stick sock out of the 8" by 14" piece of canvas. Slide the pugil stick sock over the rolled foam pad and secure it with the duct tape or riggers tape. Duct tape or riggers tape protects the edges of the centerguard and reinforces the gluing effects of the spray adhesive.

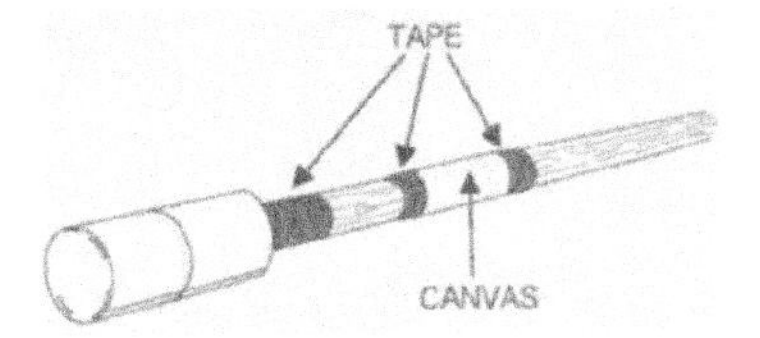

Note: It is recommended that the Fabric Repair Shop be used to make the pugil stick sock.

Finishing the Opposite End Cap

After one end cap is complete and the centerguard is finished, then repeat steps 1 through 7 in Constructing Ends, End Plugs, and End Caps for the other end of the pugil stick. Ensure that the opposite color end cap is used.

Fastening Handguards

Step 1. Place the inner handguards (longer piece) around the dowel. The inner handguard is placed on first because it protects the wrist. One inner handguard is needed for the left hand and one for the right hand.

Step 2. Fold the inner handguard around the pugil stick and run the para-cord or 5-50 cord through the metal grommets.

Step 3. Tighten down the corner ends of the inner handguards by pulling on the para-cord or 5-50 cord.

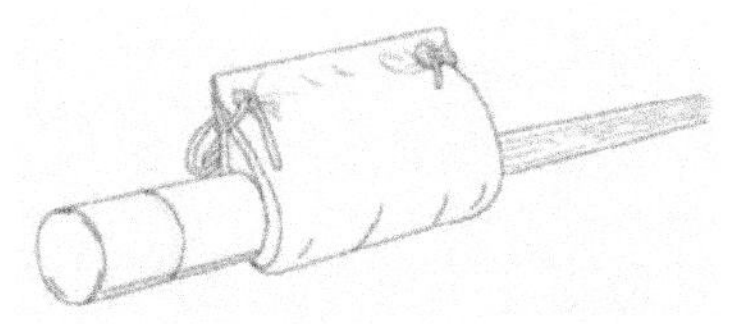

Step 4. Tie off the cord with a square knot above the metal grommets.

Step 5. Place the outer handguards (shorter piece) over the inner handguards (longer piece). Repeat steps 1-4 for the other handguard. The outer handguards provide additional protection to the knuckles and fingers and provides additional stability for the inner handguards.

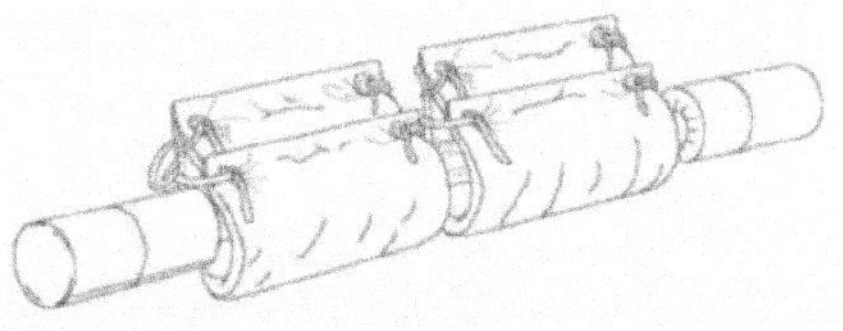

Serviceability Inspection

Prior to any training, all pugil sticks are inspected to ensure they are fit for use. If any of the following conditions exist, the pugil stick is unsafe and will not be used:

Stick protrudes through the end caps.
Frayed or worn edges exist on canvas surfaces.
The stick can be felt through the pads or the padding is too soft.
End caps or handguards are not securely attached.

APPENDIX B

SAFETY PRECAUTIONS DURING TRAINING

1. General Safety Precautions

When training close combat techniques, certain safety precautions must be adhered to to prevent injuries.

Most training should be conducted on a training area with soft footing such as a sandy or grassy area. If training mats are available, they should be used. A hard surface area is not appropriate for close combat training.

All techniques should be executed slowly at first. Marines can increase the speed of execution as they become more proficient. Marine-on-Marine training that requires contact (chokes, throws, ground fighting, and unarmed restraints and manipulation) should not be executed at full force or full speed.

If a technique is applied to the point that a Marine is uncomfortable, the Marine must "tap out." This indicates immediate release of the pressure being applied or to immediately stop the technique. The Marine "taps out" by firmly tapping his hand several times on any part of the opponent's body that will get his attention or by saying *stop*.

Second Impact Syndrome occurs when a second concussion develops within hours, days, or weeks following a prior concussion (and before recovery from the first concussion). Second Impact Syndrome causes rapid brain swelling and can cause death. Marines who experience headaches or other symptoms following training must be examined by appropriate medical personnel. These symptoms can include, but are not limited to, blurred vision, ringing in the ears, dialation of the pupils, bleeding from the ears or mouth, slurred speech, swelling in head or neck area, or any unnatural discoloration of head or neck. They should not be allowed to participate in pugil stick training or any other activity where a heavy blow might be sustained for a minimum of 7 days after the symptoms have subsided.

2. Safety Precautions for Individual Techniques

Falls

When training Marines to fall, they should progress from the ground, to a kneeling or squatting position, and then to a standing position. This ensures they are comfortable and understand the technique before progressing to executing falls from a higher profile. This instructional technique vastly reduces the risk of injury.

Strikes and Punches

When training Marines to strike and punch, they begin by executing the techniques "in the air." As they become more proficient, they execute strikes on equipment (when available) such as an air shield or a heavy bag. At no time should they be permitted to execute strikes on another student.

Chokes and Ground Fighting Chokes

When training Marines to execute chokes, they will not apply pressure to the opponent's throat during training because the trachea and windpipe can be crushed. During training, Marines should practice the proper procedures for blood chokes. They should never execute air chokes. At no time during training should a choke be applied at full force or full speed. No choke should be held for longer than 5 seconds.

Unarmed Restraints and Manipulation

When training Marines to execute unarmed restraints and manipulation techniques, they utilize slow and steady pressure. Never apply these techniques at full force or at full speed.

Pugil Stick Training

A blow to the head during training will bruise the brain. A second blow to the head can cause death. This is known as the Second Impact Syndrome. There must be 7 days between pugil stick training to prevent injury or death.

Field Manual
No. 3-34.2

Headquarters
Department of the Army
Washington, DC, [pending date]

Combined-Arms Breaching Operations

Contents

Distribution Restriction: Approved for public release; distribution is unlimited.

*This publication supersedes Field Manuals (FMs) 5-101, 23 January 1985, and 90-13-1, 28 February 1991.

Preface

FM 3-34.2 provides combined-arms commanders, from company to brigade, and their staffs with the doctrine, tactics, and techniques needed to successfully overcome obstacles. It provides the commander with breaching fundamentals and information for planning, preparing, and executing breaching operations. This manual gives the coordinating and special staff officer a basis for synchronizing all combat multipliers during a breaching operation.

This is not a stand-alone manual. The user must have a fundamental understanding of the concepts outlined in FMs 5-71-2, 5-71-3, 5-71-100, 17-98, 20-32, 34-130, 71-1, 71-2, 71-3, 100-5, 100-7, 101-5, and 101-5-1. This manual also implements Standardization Agreement (STANAG) 2036.

Appendix A contains a metric conversion chart.

The proponent for this publication is HQ, TRADOC. Send comments and recommendations on Department of the Army (DA) Form 2028 directly to Commandant, United States (US) Army Engineer School, ATTN: ATSE-DOT-DD, Directorate of Training, 320 Engineer Loop, Suite 336, Fort Leonard Wood, Missouri 65473-8929.

Unless this publication states otherwise, masculine nouns and pronouns do not refer exclusively to men.

Chapter 1

Obstacle Breaching Theory

Breaching operations are conducted to allow maneuver despite the presence of obstacles. Obstacle breaching is the employment of a combination of tactics and techniques to advance an attacking force to the far side of an obstacle that is covered by fire. It is perhaps the single, most difficult combat task a force can encounter. Understanding breaching theory is the first step to understanding breaching tactics. Breaching is a synchronized combined-arms operation under the control of a maneuver commander. Breaching operations begin when friendly forces detect an obstacle and begin to apply the breaching fundamentals, and they end when battle handover has occurred between follow-on forces and a unit conducting the breaching operation. *Bulling through* or *forcing through* is not a breaching operation. Bulling through is a decision made when a commander must react immediately to extricate his force from an untenable position within an obstacle and no other breaching operations are possible. When a force is in a minefield receiving fires and taking heavy losses, the commander may decide to immediately bull through the minefield rather than withdraw or reduce the obstacle.

DEFINITIONS

1-1. The following definitions are inherent to obstacle breaching and are used throughout this manual:

- **Obstacle**. An obstacle is any obstruction that is designed or employed to disrupt, fix, turn, or block the movement of an opposing force (OPFOR) and to impose additional losses in personnel, time, and equipment on the OPFOR. Obstacles can exist naturally (existing), be man-made (reinforcing), or be a combination of both. A complex obstacle is a combination of different types of individual obstacles that requires more than one reduction technique (explosive, mechanical, manual) to create a lane through the obstacle. A reinforcing obstacle is an obstacle that is specifically constructed, emplaced, or detonated through military effort. (For more information on obstacle classification, see *FM 90-7*.) The types of reinforcing obstacles are—
  - **Tactical**. A tactical obstacle is employed to disrupt enemy formations, turn them into a desired area, fix them in position under direct and indirect fires, or block their penetration while multiplying the effects and capabilities of firepower.
  - **Protective**. A protective obstacle is employed to assist a unit in its local, close-in protection.

- **Lane**. A lane is a route through, over, or around an enemy or friendly obstacle that provides safe passage of a passing force. The route may be reduced and proofed as part of a breaching operation, be constructed as part of the obstacle, or be marked as a bypass.
- **Reduction**. This is a task to create and mark lanes through, over, or around an obstacle to allow the attacking force to accomplish its mission. Normally, engineers and reduction assets are used to reduce an obstacle.
- **Proofing**. Proofing verifies that a lane is free of mines and that the width and trafficability of the point of breach are suitable for the assault force. Proofing can be conducted visually (against surface-laid minefields), electronically (mine detectors), or mechanically (mine-clearing rollers [MCRs]). Proofing is conducted when the risk of live mines remaining in the lane exceeds the risk of loss (lives and equipment) to enemy fires while waiting to complete proofing. Some mines are resistant to some reduction assets; for example, magnetically fused mines may be resistant to a mine-clearing line charge (MICLIC). Proofing is vitally important, considering the vast variety of mines in use. It should always be planned for a breaching operation; however, the time available, the threat, or the mission may dictate that proofing not be done.
- **Clearing**. This is the total elimination or neutralization of an obstacle or a portion of the obstacle. Clearing operations are not conducted under fire and are usually performed by follow-on engineer forces.
- **Breach area**. The breach area is established and fully defined by the higher headquarters (HQ) of the unit conducting breaching operations. It is the area where a breaching operation occurs. The area must be large enough to allow the attacking unit to deploy its support force and extend far enough on the far side of the obstacle to allow follow-on forces to deploy before leaving the breach area. One technique is to establish the breach area using phase lines (PLs) or unit boundaries. The PL defining the far side of the breach area may be established as a battle handover line (BHL).
- **Farside objective**. The farside objective is the immediate objective of the attacking unit's assault force, and it may be oriented by the terrain or the enemy force. The higher HQ assigns the objective; however, the attacking unit normally subdivides the objective into smaller objectives to assign responsibilities and to control and focus the assault of subordinate forces. When breaching as part of a larger force, seizing the farside objective provides the necessary maneuver space for the higher unit's follow-on forces to move securely through the lanes, assemble or deploy, and continue the attack without enemy interference.
- **Bypass**. A bypass is a tactical task that involves maneuvering around an obstacle, a position, or an enemy force to maintain the momentum of advance. When a unit bypasses an obstacle, it physically changes the direction of movement to avoid the obstacle. This must be done with caution because it might play into the enemy's hand. Bypassed obstacles and enemy forces are reported to higher HQ.

- **Point of penetration**. This is the location, which is identified on the ground, where the commander concentrates his efforts at the enemy's weakest point to seize a foothold on the farside objective. This is achieved along a narrow front through maneuver and direct and indirect fires that are accurately placed against enemy forces. A commander conducting a breaching operation establishes a point of penetration that supports planning locations for the reduction area and the seizure of the farside objective.
- **Reduction area**. This is a number of adjacent points of breach that are under the control of the breaching commander. The commander conducting the attack determines the size and location of the reduction area that supports the seizure of a point of penetration (see *Figure 1-1*).
- **Point of breach**. This is the location at an obstacle where the creation of a lane is being attempted. Initially, points of breach are planned locations only. Normally, the breach force determines the actual points of breach during the breaching operation.

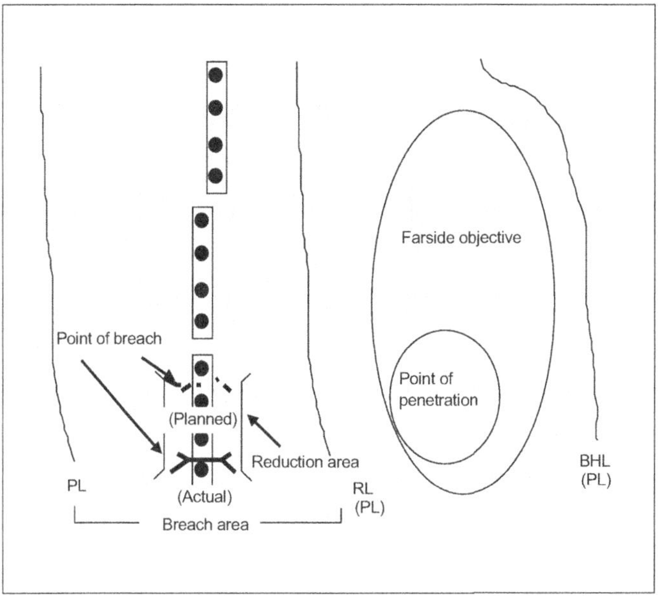

Figure 1-1. Breach Area

BREACHING TENETS

1-2. Successful breaching operations are characterized by applying breaching tenets. These tenets should be applied whenever an obstacle is encountered in the area of operations (AO), whether during an attack or during a route clearance operation. The tenets are—

- Intelligence.
- Breaching fundamentals.
- Breaching organization.
- Mass.
- Synchronization.

INTELLIGENCE

1-3. Success depends largely on the force commander's ability to *see the battle space*. He must identify how the enemy is using the terrain to minimize the risk of surprise. This is particularly true when attempting to counter the enemy's use of obstacles. The force commander does this with the intelligence preparation of the battlefield (IPB) process. During the IPB process, the situation template (SITEMP) is developed. The SITEMP is a graphic depiction of expected threat dispositions based on threat doctrine and the effects of the battle space for a particular course of action (COA). Based on the commander's estimates, the staff develops a SITEMP for multiple enemy COAs. The SITEMP helps drive the maneuver planning in support of war gaming and the development of the event template. The SITEMP also drives the decisions the commander must be prepared to make. The commander uses the decision-support template (DST) to convey these decisions in graphic form. The IPB process helps achieve success by focusing available intelligence-gathering assets toward answering the commander's priority intelligence requirements (PIR). Intelligence that is gathered later by recon becomes the foundation for developing and revising a SITEMP. (See *Appendix B* for information on threat mine operations.)

1-4. Time-constrained planning requires the rapid development of the enemy's SITEMP. A minimal SITEMP, complete with templated obstacles, allows the development of an initial event template and provides initial PIR for the intelligence-collection plan (ICP). At a minimum, the unit can template an obstacle trace that is precise, as far as the location, and represents the total expected enemy countermobility effort but does not depict individual obstacles. The staff plots the actual composition of the trace based on incoming obstacle reports. Adequate time for intelligence collection is critical in developing an accurate picture of the battle space, yet time is normally in short supply. While templating is essential to focus the ICP, too much time can be spent on templating at the expense of intelligence, surveillance, and reconnaissance (ISR).

1-5. In any operation where enemy obstacles can interfere with friendly maneuver, obstacle intelligence (OBSTINTEL) is always one of the information requirements (IR) and should become PIR. Finding enemy obstacles or seeing enemy obstacle activity validates and refines the Intelligence Officer's (US Army) (S2's) picture of the battle space.

OBSTINTEL helps the S2 determine the enemy's intentions, plans, and defense strength. The force engineer is the unit's expert on enemy countermobility capabilities. He assists the S2 in templating enemy obstacles, analyzing OBSTINTEL, and analyzing the effects of terrain.

1-6. An unverified enemy template can lead to disaster because the force may aim an attack at the wrong place. Units may deploy to reduce expected obstacles early, wasting mission time to *feel* their way into nonexistent obstacles; or they may blunder into an unexpected obstacle or an enemy engagement area (EA).

1-7. OBSTINTEL can be a critical indicator to verify an enemy template. The force's Operations and Training Officer (US Army) (S3), S2, and engineer establish effective OBSTINTEL collection by determining specific obstacle IR. Examples of information that is needed to fulfill obstacle IR include—

- The location of existing and reinforcing obstacles.
- The orientation and depth of obstacles.
- Conditions of the soil (in the case of a minefield) to determine the ability to use tank plows.
- The presence, location, and type of wire.
- Lanes and bypasses.
- The composition of the minefield (buried or surface-laid antitank [AT] and antipersonnel [AP] mines, antihandling devices [AHDs], and the depth of the mines).
- Types of mines and fuses.
- The location of enemy indirect-fire systems that can fire into the breach area.
- The composition of complex obstacles.
- Areas between successive obstacle belts.

1-8. OBSTINTEL is particularly important for discovering the types of mines and mine fuses the enemy has employed. The engineer depends on this information since he must determine which reduction techniques offer the best chance for success and minimize the risk to the breach force. This may require a dismounted recon patrol to examine mines within a minefield. The enemy may employ security patrols, have forces overwatching the obstacle, or place forces within the obstacle to prevent obstacle recon.

1-9. Obtaining OBSTINTEL requires dedicated collection assets. These assets identify survivability positions and obstacle emplacement activity. Aviation units, who are tasked to perform recon in the area, may be able to provide valuable OBSTINTEL information. Ground-surveillance radar may also provide useful information.

1-10. Recon forces should have the capability to collect OBSTINTEL. To be successful in this collection, the obstacle recon efforts must be tied into the maneuver force's overall reconnaissance and surveillance (R&S) plan. A force ensures this capability by training dedicated recon assets on collecting required obstacle information or by augmenting recon forces with engineer recon teams provided by the supporting engineer force. An engineer recon team moves with scouts or a patrol and conducts dismounted recon of templated or confirmed obstacles. This team has the flexibility to dismount and develop the details of the obstacle. Personnel engaged in a recon mission for OBSTINTEL should be used rarely, if ever, to reduce obstacles during the mission; although they make ideal guides to breach forces for subsequent breaching operations. Inadvertent detonation during reduction may compromise engineers and scouts, defeating the recon mission. It may also compromise the entire attack. *FM 5-170* describes capabilities, limitations, and employment concepts of engineer recon teams.

BREACHING FUNDAMENTALS

1-11. Suppress, obscure, secure, reduce, and assault (SOSRA) are the breaching fundamentals that must be applied to ensure success when breaching against a defending enemy. These fundamentals will always apply, but they may vary based on the specific battle-space situation (mission, enemy, terrain, troops, time available, and civilian considerations [METT-TC]).

Suppress

1-12. Suppression is a tactical task used to employ direct or indirect fires or an electronic attack on enemy personnel, weapons, or equipment to prevent or degrade enemy fires and observation of friendly forces. The purpose of suppression during breaching operations is to protect forces reducing and maneuvering through an obstacle.

1-13. Effective suppression is a mission-critical task performed during any breaching operation. Suppressive fires in sufficient volume serve to secure the reduction area. Successful suppression generally triggers the rest of the actions at the obstacle. Fire control measures ensure that all fires are synchronized with other actions at the obstacle. Although suppressing the enemy overwatching the obstacle is the mission of the support force, the breach force should be able to provide additional suppression against an enemy that the support force cannot effectively suppress.

Obscure

1-14. Obscuration must be employed to protect forces conducting obstacle reduction and the passage of assault forces. Obscuration hampers enemy observation and target acquisition and conceals friendly activities and movement. Obscuration smoke deployed on or near the enemy's position minimizes its vision. Screening smoke employed between the reduction area and the enemy conceals movement and reduction activities. It also degrades enemy ground and aerial observations. Obscuration must be carefully planned to provide maximum degradation of enemy observation and fires, but it must not significantly degrade friendly fires and control.

Secure

1-15. Friendly forces secure the reduction area to prevent the enemy from interfering with obstacle reduction and the passage of the assault force through the lanes created during the reduction. Security must be effective against outposts and fighting positions near the obstacle and against overwatching units, as necessary. The far side of the obstacle must be secured by fires or be occupied before attempting any effort to reduce the obstacle. The attacking unit's higher HQ has the responsibility to isolate the breach area by fixing adjacent units, attacking enemy reserves in depth, and providing counterfire support.

1-16. Identifying the extent of the enemy's defenses is critical before selecting the appropriate technique to secure the point of breach. If the enemy controls the point of breach and cannot be adequately suppressed, the force must secure the point of breach before it can reduce the obstacle.

1-17. The breach force must be resourced with enough maneuver assets to provide local security against the forces that the support force cannot sufficiently engage. Elements within the breach force that secure the reduction area may also be used to suppress the enemy once reduction is complete.

Reduce

1-18. Reduction is the creation of lanes through or over an obstacle to allow an attacking force to pass. The number and width of lanes created varies with the enemy situation, the assault force's size and composition, and the scheme of maneuver. The lanes must allow the assault force to rapidly pass through the obstacle. The breach force will reduce, proof (if required), mark, and report lane locations and the lane-marking method to higher HQ. Follow-on units will further reduce or clear the obstacle when required. Reduction cannot be accomplished until effective suppression and obscuration are in place, the obstacle has been identified, and the point of breach is secure. Obstacle-reduction techniques are discussed in more detail in *Appendix C*.

Assault

1-19. A breaching operation is not complete until—

- Friendly forces have assaulted to destroy the enemy on the far side of the obstacle that is capable of placing or observing direct and indirect fires on the reduction area.
- Battle handover with follow-on forces has occurred, unless no battle handover is planned.

BREACHING ORGANIZATION

1-20. A commander organizes friendly forces to accomplish the breaching fundamentals quickly and effectively. This requires him to organize support, breach, and assault forces with the necessary assets to accomplish their roles (see *Table 1-1*, page 1-8).

Table 1-1. Relationship Between Breaching Organization and Breaching Fundamentals

Breaching Organization	Breaching Fundamentals	Responsibilities
Support force	Suppress Obscure	Suppress enemy direct-fire systems covering the reduction area. Control obscuring smoke. Prevent enemy forces from repositioning or counterattacking to place direct fires on the breach force.
Breach force	Suppress (provides additional suppression) Obscure (provides additional obscuration in the reduction area) Secure (provides local security) Reduce	Create and mark the necessary lanes in an obstacle. Secure the nearside and far side of an obstacle. Defeat forces that can place immediate direct fires on the reduction area. Report the lane status/location.
Assault force	Assault Suppress (if necessary)	Destroy the enemy on the far side of an obstacle that is capable of placing direct fires on the reduction area. Assist the support force with suppression if the enemy is not effectively suppressed. Be prepared to breach follow-on and/or protective obstacles after passing through the reduction area.

Support Force

1-21. The support force's primary responsibility is to eliminate the enemy's ability to interfere with a breaching operation. It must—

- Isolate the reduction area with fires.
- Mass and control direct and indirect fires to suppress the enemy and to neutralize any weapons that are able to bring fires on the breach force.
- Control obscuring smoke to prevent enemy-observed direct and indirect fires.

1-22. Suppression is critical for a successful breaching operation; therefore, allocation of adequate forces to the support force is essential. The commander must allocate enough direct- and indirect-fire systems to the support force to allow it to accomplish its missions. When resourcing the support force, consider the losses it will receive as it fights its way into its support-by-fire (SBF) position. To increase the survivability of the support force, the commander may establish critical friendly zones (CFZs) in support of the support force once it has occupied its SBF positions. A CFZ is an area, usually a friendly unit or location, that the maneuver commander designates as critical to protect an asset whose loss would seriously jeopardize the mission. Covered by a radar sector, the CFZ supports counterfire operations by providing the most responsive submission of targets to the fire-support system when rounds impact inside the CFZ. (For more detailed information on CFZs, see *FM 6-121*.)

1-23. The support force must seize an SBF position, seeking maximum protection from the terrain. If possible, the support force should follow a covered or concealed route to the SBF position, take up its assigned sectors of fire and observation, and begin to engage the enemy. It may have to make some adjustments to its direct-fire plan if the situation dictates. Observation is particularly critical. Artillery observers with the support force may initially bring indirect fires on enemy positions to fix and suppress the enemy. The support force adjusts the artillery-delivered obscuring smoke, if necessary, to protect the breach and assault forces as they approach the reduction area.

Breach Force

1-24. The breach force assists in the passage of the assault force by creating, proofing (if necessary), and marking lanes. (For more information on the initial lane-marking pattern, see *Chapter 3*.) The breach force is a combined-arms force. It includes engineers, reduction assets, and enough maneuver forces to provide additional suppression and local security. The breach force may apply portions of the following breaching fundamentals as it reduces an obstacle:

- **Suppress**. The breach force must be allocated enough maneuver forces to provide additional suppression against various threats, including—
  - Enemy direct-fire systems that cannot be effectively observed and suppressed by the support force due to the terrain or the masking of the support force's fires by the breach force as it moves forward to reduce the obstacle.
  - Counterattacking and/or repositioning forces that cannot be engaged by the support force.
- **Obscure**. The breach force employs vehicle-mounted smoke systems and smoke pots, if necessary, for self-defense and to cover lanes while the assault force is passing.
- **Secure**. The breach force secures itself from dismounted threat forces who are providing close-in protection of the obstacle. The breach force also secures the lanes through the tactical obstacles, once they are created, to allow safe passage of the assault force.
- **Reduce**. The breach force performs its primary mission of reducing the obstacle. To support the development of a plan to reduce the obstacle, the composition of the obstacle system must be an IR.
- **Assault**. The breach force assaults through the point of breach to the far side of an obstacle and seizes the reduction area.

1-25. The breach force is comprised of two subordinate elements—security and reduction:

- **Security element**. The security element is comprised primarily of maneuver forces. It is normally responsible for providing additional suppression, obscuration, and local security.
- **Reduction element**. The reduction element is comprised primarily of reduction assets (mine-clearing blades [MCBs], MICLICs, armored vehicle-launched bridges [AVLBs], armored combat earthmovers

[ACEs], dismounted engineers, and so forth). It is normally responsible for reducing the obstacle.

Chapter 2 provides more details on the roles, responsibilities, and command and control (C^2) of the security and reduction elements.

1-26. The breach force must be able to deploy and begin reducing the obstacle as soon as enemy fires have been suppressed and effective obscuration is achieved. It can expect enemy artillery fires within a matter of minutes. CFZs should be activated at the point of breach before the commitment of the breach force.

1-27. Engineers within the breach force are allocated with the reduction assets necessary to clear mines, nonexplosive obstacles, and small gaps. Scouts, engineers, or other recon assets are also capable of finding local bypasses or existing lanes in the obstacle system.

1-28. The commander allocates engineers and reduction assets based on the number of lanes required, the terrain, and the type of obstacle. The breach force must be able to create a minimum of one lane for each assaulting company or two lanes for the assaulting task force (TF).

1-29. The commander should plan for a 50 percent loss of mobility assets in close combat. As a planning factor, he should allocate one engineer platoon with the necessary reduction assets to create and mark one lane.

1-30. After the breach force has reduced the obstacle and passed the assault force through, the breach force may be required to hand over the lane to follow-on units. At a minimum, the lanes must be marked and their locations and conditions reported to higher HQ and follow-on units as prescribed in the unit's standing operating procedures (SOP).

Assault Force

1-31. The assault force's primary mission is to destroy the enemy and seize terrain on the far side of the obstacle to prevent the enemy from placing direct fires on the created lanes. The assault force may be tasked to assist the support force with suppression while the breach force reduces the obstacle.

1-32. The assault force must be sufficient in size to seize the point of penetration. Combat power is allocated to the assault force to achieve a minimum 3:1 ratio on the point of penetration. In a breaching operation, particularly TF and above, the assault force normally maneuvers as a separate force attacking through a reduced obstacle. However, breach and assault assets may maneuver as a single force when conducting lower-level breaching operations, such as an independent company team conducting an attack.

1-33. If the obstacle is defended by a small enemy force, assault and breach forces' missions may be combined. This simplifies C^2 and provides more immediate combat power for security and suppression.

1-34. Fire control measures are essential, since support and breach forces may be firing on the enemy when the assault force is committed. Suppression of overwatching enemy positions must continue and other enemy forces must remain fixed by fires until the enemy has been destroyed. The assault force must assume control for direct fires on the assault objective as support and breach forces' fires are lifted or shifted.

[ASS

1-35. Breaching is conducted by rapidly applying concentrated efforts at a point to reduce the obstacle and penetrate the defense. Massed combat power is directed against the enemy's weakness. The location selected for breaching depends largely on the weakness in the enemy's defense, where its covering fires are minimized. If friendly forces cannot find a natural weakness, they create one by fixing the majority of the enemy force and isolating a small portion of it for attack.

Support Force

1-36. The support force suppresses the enemy to eliminate effective fires on the breach force. Smoke and terrain are used to help isolate the force under attack. Suppression requires the commander to mass enough overwatching direct fires to protect the breach force.

1-37. Normally, the TF isolates and destroys platoon-size elements in succession, starting with the platoon identified as the easiest to overwhelm. In *Figure 1-2*, the TF commander employs a tank company team to fix two defending platoons and uses screening smoke (target AA1102) to help isolate the platoon selected for attack. Objective (OBJ) Cowboy has been identified by the TF commander as the point of penetration. The center and eastern platoons are suppressed by direct and indirect fires from a second tank company team.

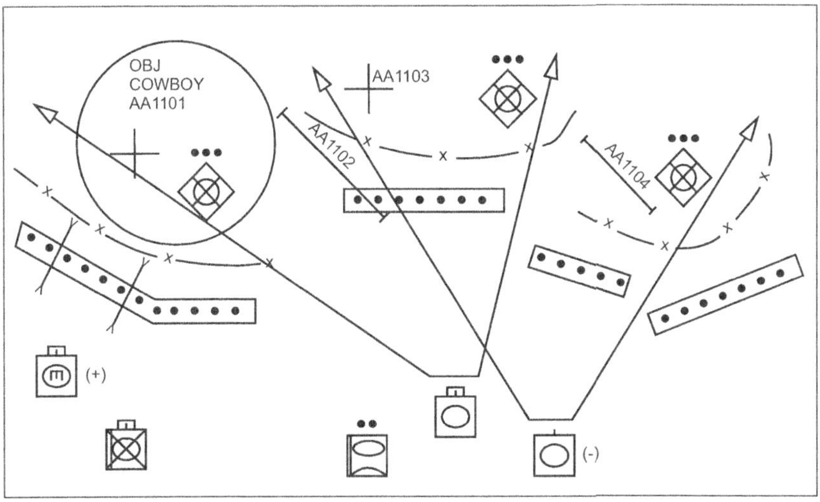

Figure 1-2. Technique for Achieving Mass in the Breach

1-38. The support force should be provided with reduction assets to reduce the impact that unexpected obstacles or scatterable minefields have on their approach to and occupation of SBF positions. Failure to do so may greatly impact on synchronizing the entire breaching operation. As a technique, a unit may create a mobility reserve that supports the main effort throughout the operation. Initially, the mobility reserve may support the support force until it has seized its SBF positions and then shift its support to the breach or assault force.

Breach Force

1-39. The commander also masses engineers and reduction assets to reduce the obstacle. The breach force is organized and equipped to use several different reduction techniques in case the primary technique fails; for example, if a key vehicle is destroyed or if casualties render dismounted engineers ineffective. Additional reduction assets are present to handle the unexpected. Normally, 50 percent more reduction assets than required for obstacle reduction are positioned with the breach force.

1-40. The breach force masses reduction efforts against the obstacle to ensure that it will successfully create enough lanes. A mounted TF-size assault force requires at least two lanes, but more lanes will speed passage through the danger area. The breach force attempts to simultaneously create as many lanes as necessary to ensure the rapid passage of the assault force through the obstacle system. This may require the breach force to simultaneously use a combination of mechanical, explosive, and manual techniques. For example, an engineer platoon may use two MICLICs as the primary technique to reduce a minefield. As soon as the engineer platoon fires the second MICLIC, engineer squads employ manual explosive techniques to create three additional lanes through the minefield. The commander must carefully plan and synchronize the creation of additional lanes to reduce the potential for fratricide with assaulting troops. The distance between the lanes depends on the enemy, the terrain, the need to minimize the effects of enemy artillery, the direct-fire plan of the support force, C^2, and reduction-site congestion.

Assault Force

1-41. Achieving necessary mass for the assault requires the breach force to open enough lanes through the obstacle to permit rapid passage and the buildup of forces on the farside. The size of the assault force determines the number of lanes initially created. A mounted TF requires one lane per assaulting company team. A dismounted assault force requires one lane for each leading assault platoon. The tactical situation may require additional lanes to quickly pass a large assault force through the obstacle to achieve sufficient combat-power ratio.

1-42. The principle of mass influences the selection of the point of breach; the task organization of the support, breach, and assault forces; and the integration of engineers in movement or attack formations. *Figure 1-2*, page 1-11, shows a technique to achieve mass in the breach.

1-43. The need to generate enough mass and the maneuver space available strongly influence which echelon can conduct the breaching operation. A company team cannot simultaneously mass sufficient fires, reduce the obstacle, and assault the defending position unless it is a simple obstacle defended by no more than one platoon. A TF should have a combat-power ratio of 3:1 to reduce an obstacle defended by a company, and a TF is normally the echelon used to execute a breach. A brigade has sufficient combat power to attack a complex, well-defended obstacle; but a brigade has difficulty deploying all its combat power within range. Normally, a brigade breaches by isolating a small segment of the defense (platoon or company) that the TF can then attack as the breaching echelon. If obstacles and defenses are in-depth, mass is achieved by passing additional TFs through the obstacle to continue the attack.

SYNCHRONIZATION

1-44. Breaching operations require precise synchronization of the breaching fundamentals by support, breach, and assault forces. Failure to synchronize effective suppression and obscuration with obstacle reduction and assault can result in rapid, devastating losses of friendly troops in the obstacle or the enemy's EA.

1-45. A combined-arms breach is a complex operation by nature. Support, breach, and assault forces must apply the SOSRA breaching fundamentals within a short time and distance. The support force masses its direct fires and controls indirect fires in concert with breach and assault forces' maneuvers. The commander must employ smoke at the right time and place to maximize its effectiveness or risk hampering his own target acquisition and C^2. The breach force must have the right reduction asset for the type of obstacle encountered. Engineers must ensure that they do not prematurely exhaust reduction assets needed to reduce subsequent obstacles. *Table 1-2*, page 1-14, shows the complexity of a combined-arms breach.

1-46. The commander ensures synchronization through proper planning and force preparation. Fundamentals to achieve synchronization are—

- Detailed reverse planning.
- Clear subunit instructions.
- Effective C^2.
- Well-rehearsed forces.

Detailed Reverse Planning

1-47. Synchronizing a combined-arms breach begins by using the reverse-planning process to ensure that actions at obstacles support actions on the objective. Planning the breach without regard to actions on the objective leads to disaster. During COA development, the staff analyzes the relative combat power and compares enemy and friendly strengths and weaknesses. The staff then generates options based on the commander's guidance and arrays the forces necessary to accomplish the mission. The staff develops the options by analyzing the objective and allocating forces to accomplish the mission. The commander decides how he must attack the objective to accomplish his mission. This decision drives where, how, and with what force he must support, breach, and assault through the enemy's obstacles. The commander approves the scheme of maneuver for the breaching phase of the operation

Table 1-2. Breach Complexity

Action	Element	Time (Minutes)	Controlled By
Develop the situation (verify the boundary of the enemy obstacle system).	Force in contact	M to 2	S3
Maneuver the support force into the overwatch position.	Support	M + 2 to 15	Support cdr
Maneuver the assault force into the covered assault position.	Assault	M + 2 to 15	Assault cdr
Call for artillery.	DS artillery	M + 2 to 15	FSO
Build smoke.	Mortars	M + 5 to 10	FSO
Suppress the enemy with direct fires.	Support	M + 15 to 29	Support cdr
Suppress the enemy with artillery fires.	DS artillery	M + 10 to 29	FSO
Maintain smoke.	DS artillery/mortars	M + 10 to 30	FSO
Maneuver the breach force to the breach location.	Breach	M + 20 to 23	Reduction cdr
Reduce the obstacle, and prepare two lanes.	Breach	M + 23 to 30	Engineer ldr
Place smoke pots.	Breach	M + 23 to EOM	Reduction cdr
Shift direct fires off the objective.	Support	M + 29 to 30	Assault cdr
Shift indirect fires beyond the objective.	DS artillery	M + 29 to 30	Assault cdr
Assault to destroy the enemy on the far side of the obstacle.	Assault	M + 30 to 45	Assault cdr
Reorganize to continue the mission.	TF	M + 45 to EOM	S3
NOTE: M = Contact with the obstacle			

that achieves adequate suppression, obscuration, and security. He also creates sufficient lanes to rapidly project combat power on the point of penetration and the far side of the objective, not just to the far side of the obstacle. Reverse planning gives purpose to the breaching plan that supports mission accomplishment.

1-48. The commander begins developing COAs as discussed in *FM 101-5*. He determines relative combat-power ratios, generates options, and arrays forces to accomplish the mission. He then develops the scheme of maneuver, including actions on the objective. In *Figure 1-3*, a three-company-team TF is conducting a deliberate attack to destroy a motorized infantry company (MIC) in the defense. The commander believes seizing OBJ Cat to be decisive. This is the point from which reverse planning of actions on the objective begins. The commander uses Team C to seize the initial foothold (OBJ Bird) (point of penetration) into the enemy's defense. The focus of breach planning is to get Team C to its objective. The number and location of lanes are driven by Team C's maneuver on OBJ Bird as the assault force. The commander now plans how the breach force must maneuver to reduce the obstacle. Supporting fires from Company A and Team B in SBFs 15 and 25 are planned to support the reduction of the obstacle and the assault on the objective. Reverse planning continues to drive the maneuver formation to ensure that forces are in the correct relative positions to accomplish their missions.

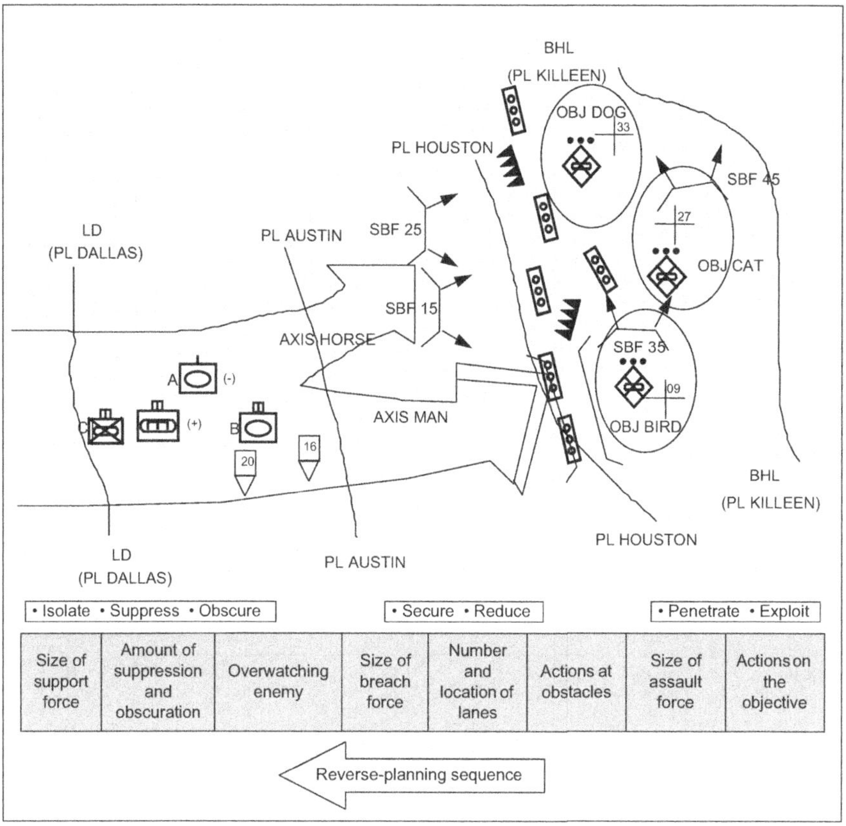

Figure 1-3. Reverse Planning

Clear Subunit Instructions

1-49. Subordinate units must clearly understand their missions, how their missions relate to the phases of the operation, and what role adjacent units play in the overall plan. The commander may use an execution matrix to synchronize his forces. An execution matrix is a superb synchronization tool, and it lists subunit instructions sequentially in relation to key events or the sequence of the attack. It also provides subordinate commanders with an understanding of how their missions and those of adjacent units fit into the overall plan. More importantly, it allows subordinates to better track the battle and coordinate their own maneuver with that of adjacent units. This is critical to achieving unity of effort between support, breach, and assault forces. *Figure 1-4* shows a sample execution matrix for the scheme of maneuver in *Figure 1-3,* page 1-15.

	Company A (Support Force)	**Team B (Support/ Assault Force)**	**Team C (Assault Force)**	**Team Engineer (Breach Force)**	
Units	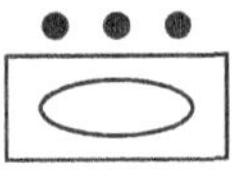	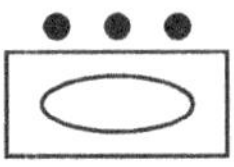	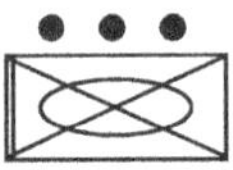	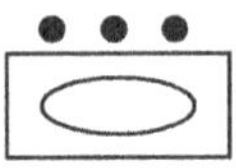	Force Allocation
	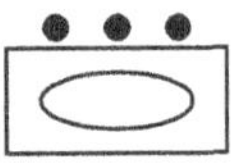	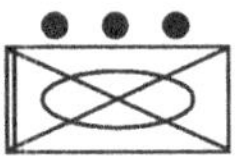	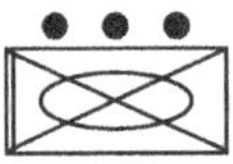	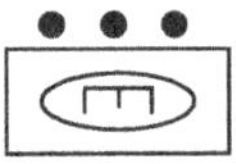	
			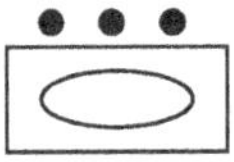	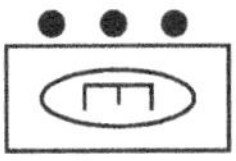	
LD to PL Austin	Move to left of Team B; echelon left.	Lead TF.	Trail team engineer.	Trail Team B center.	Maneuver formation ↓
PL Austin to PL Houston	Attack with Team B on Axis Horse. Occupy SBF 25.	Lead attack on Axis Horse. Occupy SBF 15.	Occupy CP 20.	Lead attack on Axis Man.	
Obstacle reduction	SBF 25. Orient TRP 09. On order TRP 27.	SBF 15. Orient TRP 09.	Occupy CP 16. Identify lane.	Create and mark two lanes (vicinity AB123456).	
Attack on OBJ Bird	Shift fires to TRP 33.	Shift fires to TRP 27.	On order, move through lane to seize OBJ Bird.	Assist passage of Team C. Secure lanes.	Reduce ↓
Attack on OBJ Cat	Orient TRP 33.	On order, lift fires and occupy CP 16. Identify lanes.	On order, seize OBJ Cat.	Secure/improve lanes.	
Attack on OBJ Dog	Lift fires. Consolidate on SBF 25.	On order, seize OBJ Dog.	Occupy SBF 45. Orient TRP 33. On order, lift fires.	Assist passage of Team B. Secure/improve lanes.	Action on OBJ ↓
Consolidate	SBF 25. Orient north.	OBJ Dog. Orient northeast.	OBJ Cat. Orient east.	Support hasty defense operations.	

Figure 1-4. Sample Execution Matrix

Effective C^2

1-50. Effective C^2 is paramount to mission success. C^2 is integrated into the plan through the use of maneuver and fire control measures and the positioning of key leaders to see the battle space. Maneuver control measures enable the commander to graphically convey his intent, scheme of maneuver, and subunit instructions. Relating subunit actions to the terrain is critical to successful execution.

1-51. In *Figure 1-3*, page 1-15, the commander uses PLs to initiate the breaching operation and to control actions on the objective. At PL Austin, the support force maneuvers to occupy SBFs 15 and 25. The assault force occupies checkpoint (CP) 16 to prevent premature advancement to the obstacle, reduction-site congestion, and the engagement of forces before lanes are created and marked. The commander uses target reference points (TRPs) on obvious terrain features to orient, focus, and shift suppressive direct and indirect fires.

1-52. Key leaders must be able to see the battle space to make informed decisions. Nowhere is this more true than during the breaching operation. The commander must position himself where he can best control the battle. Since effective suppression is the most critical event during breaching operations, the commander may position himself with the support force. This enables him to personally influence fire control and facilitate the necessary cross talk between breach and assault forces. The S3 may initially move with the breach force to track the progress of obstacle reduction and anticipate the commitment of the assault force. A commander who feels that his personal influence is required with the breach or assault force must make a conscious effort to track the entire battle and must not focus completely on the reduction of the obstacle or the assault itself.

Well-Rehearsed Forces

1-53. The most effective synchronization tool available to the commander is the rehearsal. The inherent complexity of the breaching operation makes rehearsals at every level essential to success. The commander must give his subordinates time to plan how they will execute their assigned missions and time to rehearse the plan with their unit. The TF facilitates rehearsals by preparing rehearsal sites and managing their use. The TF commander must be sensitive to the impact TF-level rehearsals have on subordinate planning and preparation, and he must carefully choose the time and place of the rehearsal. Company teams rehearse their support, breach, and assault roles. TF rehearsals focus on synchronizing the maneuver of support, breach, and assault forces to achieve the breaching fundamentals. Rehearsals also highlight key events that must be coordinated during breach execution.

Typical Breaching Operations

1-54. *Table 1-3*, page 1-18, shows the types of breaching operations versus the sizes of the enemy forces overwatching the obstacle.

Table 1-3. Types of Breaching Operations Versus Enemy Sizes

Enemy Overwatching the Obstacle	Maneuver Units Conducting Breaching Operations		
	Brigade	TF	Company
MIBN	X		
MIC	O	X	
MIP		O	X
NOTES: **X = Normal (combat-power ratio 3:1)** **O = Desirable (combat-power ratio >3:1)**			

BREACHING OPERATIONS IN SUPPORT OF DELIBERATE AND HASTY ATTACKS

1-55. Attacks take place along a continuum based on the knowledge of the enemy's capabilities and disposition and the intentions and details of the friendly force's planning and preparation. The terms *deliberate attack* and *hasty attack* refer to the opposite ends of that continuum and describe characteristics of the attack:

- **Deliberate attack**. Friendly forces have more knowledge of the enemy situation and have more time to plan and prepare for the enemy. With this knowledge, friendly forces conducting the deliberate attack can develop detailed plans, task-organize to accomplish the mission, and execute rehearsals down to the lowest levels.
- **Hasty attack**. The enemy situation is vague, or it changes rapidly. Friendly forces have little or no time to plan or prepare for this type of operation. Because the primary purpose of the hasty attack is to maintain momentum or take advantage of the enemy situation, it is normally conducted only with the resources that are immediately available.

1-56. Breaching operations may be required to support an attack anywhere along the continuum. During a deliberate attack, friendly forces, such as engineers and smoke platoons, are task-organized before the attack to best support the concept of the operation. In a hasty attack, there may be insufficient time to task-organize assets, such as engineers and reduction assets, in the best manner. Additionally, the unit may not have time to plan the breaching operation in detail but may have to rely on the unit's SOP for execution. Regardless of where the attack falls along the continuum, the breaching tenets apply when conducting the breaching operation in support of the attack.

PLANNING SEQUENCE

1-57. The scheme of maneuver, engineer operations, fires, air defense, and actions at obstacles are all based on the same SITEMP. Planners develop the breaching plan using the following sequence:

- Reverse planning begins with actions on the objective.

- Actions on the objective drive the size and composition of the assault force.
- The size of the assault force determines the number and location of lanes to be created.
- Lane requirements and the type of obstacle drive the amount and type of mobility assets task-organized to the breach force.
- The ability of the enemy to interfere with the reduction of the obstacle determines the size and composition of the security element in the breach force.
- The ability of the enemy to mass fires on the point of breach determines the amount of suppression required and the size and composition of the support force.

Chapter 2

Breaching Operations

For a breaching operation, a unit develops a scheme of maneuver specifically designed to cross an obstacle and continue a mission. Maneuver company teams, TFs, and brigades can conduct breaching operations. Normally, a TF executes a breaching operation and the company teams are assigned as support, breach, and assault forces. Units will conduct a breaching operation when—

- The force-allocation ratio indicates that a confirmed enemy situation is beyond the subordinate unit's capabilities to successfully conduct a breaching operation. (When companies cannot successfully conduct a breaching operation, the battalion will.)
- A subordinate unit has failed in its attempt to breach an enemy's tactical obstacles.

NOTES:
1. See *Appendix D* for a breaching scenario and special planning considerations for breaching operations in restricted terrain.
2. See *Appendix E* for information on route clearance operations.

PLANNING

2-1. Units develop plans to conduct a breaching operation using the following military decision-making process (MDMP):

Step 1. Receipt of mission.
Step 2. Mission analysis.
Step 3. COA development.
Step 4. COA analysis (war game).
Step 5. COA comparison.
Step 6. COA approval.
Step 7. Orders production.

The following paragraphs highlight key portions of the MDMP as it applies to the planning of a breaching operation. For a complete discussion of the MDMP, see *FM 101-5*.

RECEIPT OF MISSION

2-2. Before receiving a higher HQ order, a unit should receive a series of warning orders (WOs) and available intelligence products. If the products are not issued with the WOs, the higher HQ liaison officer (LO) should locate and deliver the products as soon as they become available. Intelligence products are essential for the unit to begin developing its own intelligence products and to facilitate parallel planning.

2-3. Upon receiving the higher HQ SITEMP, the unit develops standard terrain products using TerraBase II or a similar computer-software program. (For examples of TerraBase products, see *FM 5-71-3*.) This set of terrain products is called a terrain-visualization mission folder. (For more information on this subject, see *FM 5-105*.) Terrain products that support the planning of a breaching operation should be included in the unit's SOP and may include—

- Visible area plots from each of the templated enemy locations to the expected EA.
- Visible area plots from potential SBF positions and points of breach to the templated enemy location.
- Oblique view of the AO.
- Perspective view of intervisibility lines.
- Visible area plots from named areas of interest (NAIs), when NAIs are developed.

2-4. Visible area plots from each of the templated enemy locations help determine if the enemy would select these locations. For example, the enemy would not select a location directly behind a small hill because the hill would prevent the enemy from firing into a kill sack. Additionally, by combining the visible area plots on an overlay, the enemy's ability to mass fires is more easily understood. It also helps select SBF locations, obscuration locations, and sectors of fires and helps determine the breach organization for future operations.

2-5. Visible area plots from potential SBF positions and points of breach assist future planning by allowing the commander to visualize the support and breach forces' ability to suppress the enemy. Multiple plots along the length of an SBF position may be necessary to fully understand the suitability of an SBF position.

2-6. An oblique view of the AO helps the commander better visualize the terrain. A perspective view of intervisibility lines in the AO allows him to visualize how the unit can best use the terrain during a breaching operation. It also provides an idea of how the enemy may use the intervisibility lines when positioning its assets and special munitions.

2-7. Visible area plots from developed NAIs can be used to provide positioning guidance for observation posts (OPs). These products are issued to recon assets who are tasked to provide information on various NAIs.

2-8. When receiving a mission, pay particular attention to the actions the higher HQ takes in support of the breaching operation or its decisions that influence the breaching operation. Responsibilities of the higher HQ include—

- Resourcing the unit conducting the breaching operation with additional or special assets not normally task-organized to subordinate units.
- Fixing enemy forces to prevent them from repositioning and interfering in the breaching operation.

- Using scatterable mines (SCATMINEs), deep fires, attack helicopters, and close air support (CAS) to isolate the farside objective and prevent a successful enemy counterattack.
- Planning a forward passage of lines by follow-on forces through the unit conducting the breaching operation. This must include a plan to upgrade/maintain lanes through the obstacle and to create additional lanes, if necessary, for the passage of follow-on forces.
- Providing criteria to transition to its own breaching operation.

MISSION ANALYSIS

2-9. During mission analysis, the commander conducts the IPB. A product of the IPB process is the SITEMP. The SITEMP is a graphic depiction of expected threat disposition if it adopts a particular COA. The staff must combine higher HQ intelligence with its knowledge of the enemy to produce a SITEMP with the necessary detail to successfully plan a breaching operation. A SITEMP should include—

- Likely enemy EAs.
- The location and orientation of enemy forces.
- Counterattack objectives and the location of enemy reserve elements.
- The location and range of all direct- and indirect-fire systems (tanks, AT weapons, artillery).
- Enemy obstacle systems, including tactical and protective obstacles and SCATMINEs. The template must depict the enemy's countermobility capability.
- The enemy's use of nuclear, biological, chemical (NBC) weapons, including the ranges of delivery systems.
- The location of enemy target-acquisition assets.
- Likely air avenues of approach (AAs), including fixed and rotary-wing aircraft.
- Positioning of enemy air-defense artillery (ADA) assets.

2-10. Selecting the initial commander's critical-information requirements (CCIR) is a part of mission analysis, and it is crucial to begin planning recon operations that will result in useful intelligence, including OBSTINTEL. Typical CCIR for a breaching operation may include the—

- Location, composition, and orientation of the obstacle and available bypasses in the vicinity of the reduction area and the point of penetration (PIR).
- Location and composition of enemy forces that are capable of employing direct and indirect fires on the point of breach (PIR).
- Maintenance status and location of all reduction assets (friendly forces information requirements [FFIR]).
- Status of commitment criteria for the breach force (PIR and FFIR).

NOTE: The initial CCIR is developed during COA analysis.

COURSE-OF-ACTION DEVELOPMENT

2-11. When developing a COA, the commander must remember that the breaching operation is only a portion of the higher HQ operation. The unit must ensure that the end state of the COA supports the higher HQ concept of the operation. As stated in *FM 101-5*, one of the criteria that each developed COA must meet is completeness. In a breaching operation, units use detailed reverse planning, develop a plan to achieve the breaching fundamentals, develop clear subunit instructions to develop a complete COA, and begin to synchronize the operation.

2-12. The detailed reverse-planning process, the SITEMP, and the terrain products produced during mission analysis are used to determine the size and composition of support, breach, and assault forces. The reverse-planning process begins with actions on the objective and includes templated enemy obstacles.

2-13. The assault force's task and purpose drive its composition. The assault force's mission is to seize terrain on the farside of the obstacle to prevent the enemy from placing direct fires on created lanes. Consider the following when determining the assault force's size and composition:

- Higher HQ requirements in terms of seizing or securing objectives.
- The terrain.
- The expected disposition, composition, and strength of the enemy when the assault force is committed.

2-14. The assault force's size determines the number and location of lanes to be created. The breach force creates and marks lanes through or around obstacles to support the rapid passage of the assault force. An assaulting company team requires a minimum of one lane, and an assaulting TF requires at least two lanes.

2-15. Lane requirements and the type of obstacle drive the amount and type of reduction assets task-organized to the breach force. As stated in *Chapter 1*, the commander should plan for the loss of about 50 percent of his reduction assets during a breaching operation. To ensure that the breach force is adequately resourced, the initial allocation of assets is continually reviewed as additional intelligence information (including OBSTINTEL) is received.

2-16. The enemy's ability to interfere with obstacle reduction determines the size and composition of the security element in the breach force. The security element's size and composition within the breach force can also be determined through the—

- Terrain analysis and the terrain's impact on the support force's ability to suppress the enemy.
- Amount of terrain the breach force will occupy at the obstacle during reduction and the extent that the breach force will mask the fires of the support force.
- Analysis of the enemy's expected reaction to the reduction effort.

2-17. Detailed terrain analysis and the enemy's expected size, disposition, and composition determine the enemy's ability to mass fires at the point of breach. That ability, in turn, determines the amount of suppression that is required and the support force's size and composition. Expected losses that the support force will incur from the line of departure (LD) to the SBF position due to maintenance and enemy air, chemical, and artillery attacks should be considered. Additionally, analyzing the enemy's equipment type, range, and lethality can indicate the expected losses that the support force will incur while in an SBF position.

2-18. When developing COAs for a breaching operation, consider the following:

- **Suppress**.
  - Identify the support force's axis of advance to SBF positions. The support force's mobility requirements must be identified based on the SITEMP and its axis of advance to SBF positions.
  - Establish SBF positions for each element in the support force.
  - Establish TRPs to control the support and breach forces' direct fires.
  - Determine the location and size of CFZs to protect the support force.
  - Plan essential indirect-fire targets to assist in suppressing the enemy.
  - Develop a plan for lifting or shifting direct and indirect fires.
  - Deploy the support force into a position to fire simultaneously on the enemy from several directions, if possible.
  - Attack by fire before or during obstacle reduction to suppress an overwatching enemy. If circumstances allow, forces may bypass the obstacle and assault the overwatching enemy before reduction. If the assault successfully eliminates the enemy, it also eliminates direct fires on the breach force. If the assault does not succeed in destroying the enemy, it fixes them and reduces the enemy's ability to place fires on the breach force. Direct- and indirect-fire suppression from the support force is necessary to support this assault.
  - Use fires to suppress the enemy's indirect fires. The enemy's indirect fires are extremely dangerous during a breaching operation since the breach force may be exposed for a lengthy period. Friendly artillery and air assets support the breaching operation by attacking enemy overwatch positions, mobile reserves, and artillery positions that are capable of firing on the point of breach. Friendly forces jam enemy fire control nets. The fire-support officer (FSO) designates points of breach as CFZs for a target-acquisition battery so that enemy artillery is treated as a priority target for counterbattery support.
- **Obscure**.
  - Determine the location of smoke target(s).
  - Identify terrain features that mask the force's point of breach, and limit the number of enemy forces that can effectively observe and fire on the breach force, thus aiding in obscuration. Other natural, limited-visibility conditions (darkness, fog) should be exploited if possible. Smoke is generally the primary obscurant; however, it is

a double-edged sword because it may attract enemy attention and degrade friendly target acquisition and C^2.

- Determine the quantity, location, and type of obscurant required to screen breach and assault forces and obscure enemy fires. This is particularly important for breaching complex obstacles where the breach force is exposed for an extended time. Smoke is important in deceiving the enemy as to the breaching location and interfering with enemy fires. If smoke platoons are available, synchronize them with artillery-delivered and organic smoke.
- Determine smoke requirements by using *FM 3-50*. Blanket, haze, or curtain-screening smoke degrades enemy observation of assault and breach forces. A smoke platoon is capable of providing support for a 0.5- to 1.4-kilometer front. Friendly forces use a combination of carefully placed smoke pots and indirect fires to augment smoke support. Howitzers and mortars supporting friendly forces can build and maintain smoke, but smoke effectiveness depends on the number and rate of smoke rounds delivered. Using indirect fire to provide smoke competes with the friendly force's high-explosive (HE) and dual-purpose improved conventional munition (DPICM) missions against an enemy. A commander must decide the priority for the use of indirect fire, and his decision must be clear to subordinates. Vehicles equipped with smoke-grenade launchers can provide local smoke for reduction operations and for troop passage during an assault. However, these systems are self-defense mechanisms and friendly forces must not rely on them to supply all obscuration for a breaching operation.

- **Secure**.
  - Plan direct-fire control measures as necessary.
  - Use terrain and the positioning of nearside and farside security SBF near the point of breach to facilitate the survivability of the breach force.
  - Plan a CFZ and ADA coverage around the reduction area for survivability and security of the breach force.
  - Plan local security at the breach site to suppress overwatching enemy forces that are close to the obstacle.
- **Reduce**.
  - Identify primary and alternate reduction techniques.
  - Determine the tentative location of lanes and the separation between lanes.
  - Determine the breach force's task organization.
  - Position the breach force, and identify positions and routes that it uses as it maneuvers toward the obstacle to minimize its impact on hindering the support force's suppressive fires.
- **Assault**.
  - Plan to pass the assault force through lanes created in the obstacle.
  - Develop objectives for the assault force.

- Develop SBF positions to be occupied by the assault force to facilitate passage of follow-on forces.
- Plan direct- and indirect-fire control measures that support the seizure of the farside objective.
- Identify all mobility requirements for the COA, including whether the assault force requires engineer support for reducing protective obstacles and fortifications within the defensive position.

2-19. Each element within the breach organization must completely understand which tasks it is responsible for in breaching operations. Additionally, each COA should identify which force is the main effort throughout the breaching operation. For example, the COA should define when the main effort shifts from the support force to the assault force. The following are the different types of forces and their tasks and responsibilities:

- **Support force**.
  - Suppress enemy elements capable of placing direct fires on the point of breach to protect the breach force.
  - Suppress enemy elements capable of placing direct fires on the assault force. This may require the adjustment of SBF positions.
  - Call for and adjust indirect fires, including obscuration. A combat observation lasing team (COLT) or Striker will be required for Copperhead munitions.
  - Fix enemy forces that are capable of repositioning.
- **Breach force**.
  - Reduce lanes in the obstacle.
  - Provide local security (farside and nearside).
  - Provide additional suppression of enemy forces overwatching the obstacle.
  - Mark and report the location of created lanes.
  - Assist the passage of the assault force through created lanes.
- **Assault force**.
  - Seize the farside objective.
  - Reduce protective obstacles.
  - Prevent enemy direct fire from interfering with follow-on forces as they pass through lanes.
  - Provide clear routes to the BHL for follow-on forces.
  - Conduct battle handover with follow-on forces.

2-20. The assault force's mission is unique. It must reduce the enemy's protective obstacles, assault the enemy's position, and secure terrain that is essential to the passage of follow-on forces. The assault-force commander must thoroughly understand the higher HQ scheme of maneuver. The following information is particularly important:

- Objectives the assault force must seize.
- The end state, in terms of SBF positions, which the assault force must occupy so that follow-on forces can continue the attack.

- The number of lanes the breach force plans to create to facilitate the passage of the assault force.
- The higher HQ plan to provide suppressive fires and/or obscuring smoke.
- Planned points of breach and the point of penetration.
- Lane requirements (number and lane-marking type) to support the passage of follow-on forces.
- The sequence of follow-on units to be passed through lanes and the criteria to begin passing follow-on units through lanes that the assault force created.

2-21. Typically, the enemy emplaces protective obstacles 50 to 500 meters in front of its positions and between its forward platoons and companies. These obstacles are designed to prevent a mounted assault and to fix or break up a dismounted assault. Close-in obstacles may be a combination of wire, AP and AT mines, fortifications, and entrenchments. All obstacles are covered by direct fire. Mines may be surface-laid or buried and well-camouflaged. Mines can be deployed with AHDs that are command-detonated or trip-wire-activated.

2-22. The assault force commander considers the following breaching fundamentals when developing his plan:

- **Suppress**. If a company team is the assault force as part of a TF breaching operation, other company teams within the TF will normally have the responsibility to suppress the enemy from designated SBF positions on the objective area. These teams may be the same support force that protected the breach force; however, terrain factors may require repositioning to provide effective support for the assault force.
- **Obscure**. Mortars and vehicle-mounted obscurants are primary means of obscuration for the assault force. It may also employ hand-held smoke grenades and smoke pots to provide additional obscuration. If available, smoke may be employed away from the point of breach to confuse enemy forces as to the disposition of the assault force.
- **Secure**. Local security must be established before conducting reduction operations on protective obstacles.
- **Reduce**.
  - The assault force's reduction efforts center on creating dismounted/mounted lanes in protective obstacles, reducing fortifications, and widening lanes to support follow-on forces. The commander should task-organize the assault force with sufficient reduction assets to create a minimum of one lane per assaulting platoon. There may be insufficient time to reorganize engineers and reduction assets to the assault force after crossing the LD. Therefore, engineers and reduction assets supporting the assault force are task-organized to the maneuver force they will support.
  - During the planning process, the assault force commander must decide whether he will assault mounted or dismounted. This decision impacts the reduction methods that will be used. A

mounted assault allows the use of tank plows and MICLICs. During a dismounted assault, friendly forces rely on manual obstacle reduction, which is very time-consuming. The composition of follow-on forces must also be considered. This may require widening initial lanes that the assault force created, especially when a dismounted assault is followed by mechanized follow-on forces.

- The assault force normally decentralizes C^2 for the plan's execution. For this reason, an assault force commander must allocate engineers, reduction assets, and demolitions to squads and platoons to provide reduction capability throughout his force.

- **Assault**. Below-ground operations require clearing the enemy's trench lines and bunker complexes. Dismounted infantry and engineers, in concert with tanks and Bradley fighting vehicles (BFVs), conduct these operations. These actions must be planned in detail and rehearsed to ensure proper execution. See *FMs 71-1* and *7-10* for more information.

COURSE-OF-ACTION ANALYSIS (WAR GAME)

2-23. When planning a breaching operation, war gaming is critical to refine the COA, including—

- Task-organizing support, breach, and assault forces.
- Developing a DST, identifying triggers to support obscuration, committing breach and assault forces, and shifting fires.
- Identifying and/or confirming the locations of decision points, NAIs, targeted areas of interest (TAIs), and other information that is needed to support decision points.
- Refining CCIR and incorporating them into the ISR plan/graphics.
- Refining the fire-support plan/graphics, including the obscuration plan.
- Confirming the obstacle template according to the TF R&S plan.

2-24. The commitment of the breach force is a critical decision point that must be included in the DST. The commander must thoroughly analyze what criteria must be met before the breach force begins obstacle reduction. Commitment criteria elements may include the following:

- Destruction of certain vehicles or a certain number of vehicles.
- Effective suppression of the enemy by the support force.
- Effective obscuration of the enemy.
- Remaining strength of the support force.
- Remaining reduction assets available to the breach force.
- Activation of CFZ.
- Air-defense assets in position.

NOTE: A plan must exist on how information will be gathered to verify the commitment criteria.

2-25. Once war gaming is complete, a COA has been refined, and a final task organization has been determined, a C^2 system can be planned to support the operation. To conduct a successful breaching operation, an effective C^2 system must be established. Specifically, each force in the breaching organization (support, breach, and assault) must understand who is in charge, especially within support and breach forces.

2-26. The support force may be comprised of more than one element. For example, in a TF breaching operation, the support force may be comprised of two company teams. Due to the importance of the support force's role and its size, it may be best to have the TF commander/S3 be the support force commander. He is responsible for coordinating the actions of both company teams and determining when the conditions have been met for the breach force's commitment. This allows each company commander to concentrate on directing his unit while under the overall control of an experienced leader.

2-27. The breach force, with its two subordinate elements (security and reduction), must have a well-understood C^2 system in place. There are many ways to establish C^2 of the breach force, and the method used may depend on the—

- Size and composition of security and reduction forces.
- Number of lanes that must be created.
- Experience level of the personnel within the breach force.

2-28. The following are possible C^2 structures within the breach force for a TF breaching operation:

- A company-team commander acts as the security element commander and the breach force commander, with an engineer as the reduction element commander.
- An engineer-company commander acts as the reduction element commander and the breach force commander, with a maneuver platoon leader as the security element commander.

2-29. Another C^2 issue that must be resolved during COA analysis is traffic control through lanes created in an obstacle system. *Figure 2-1* shows an example of the graphic control measures that may be established to facilitate the efficient movement of forces through an obstacle system. In this example, traffic-control post (TCP) 8 is located at the far-recognition marker. These control measures can be initially established by the higher HQ but need to be refined as the unit conducting the breaching operation develops its scheme of maneuver. *Table 2-1* lists the responsibilities for creating lanes and controlling traffic through an obstacle system when a TF breaching operation is being conducted.

PREPARATION

2-30. Intelligence collection continues throughout preparations for a breaching operation. As intelligence reports are received, the SITEMP and the ISR plan are updated and revised. Intelligence reports are also used to refine the task organization of support, breach, and assault forces and the scheme of maneuver, including proposed points of breach and target grids for smoke and suppressive fires. Updated intelligence information is also used during combined-arms rehearsals.

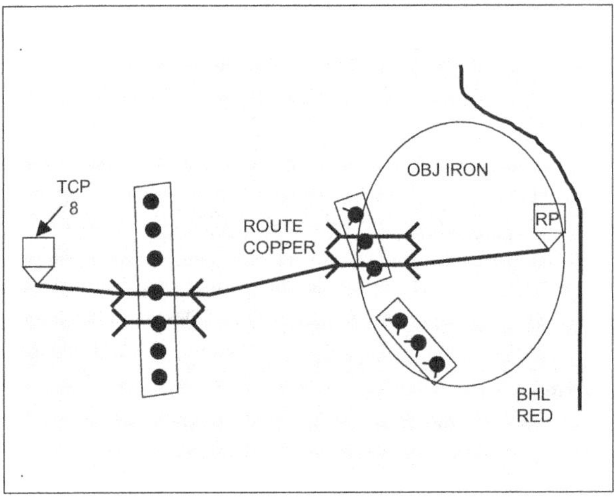

Figure 2-1. Graphics to Support Passage of Follow-On Forces

Table 2-1. C^2 Responsibilities for Lanes and Traffic Control

Breaching Operation	Responsibilities of TF Forces	Responsibilities of the Brigade
Initiating breaching operations through the passage of the assault force through tactical obstacles	The breach force creates lanes for the assault force (reduces, proofs, and emplaces the initial lane-marking pattern and assists in the passage of the assault force through the marked lanes). The TF controls all traffic passing though tactical obstacles.	
Completing the passage of the assault force through the seizure of the farside objective	The assault force creates and marks lanes through protective obstacles up to the BHL.	Assumes control of lane(s) in tactical obstacles Creates more lanes as necessary Upgrades lane marking to the intermediate lane-marking pattern Ensures that lanes remain open Assumes traffic control through tactical obstacles
Passing follow-on forces through tactical and protective obstacles to the BHL	The assault force assists in the passage of follow-on forces (from tactical obstacles to the BHL).	
Completing the forward passage of lines		Assumes control of all lanes and traffic up to the BHL

2-31. A unit meticulously plans, manages, and controls breaching rehearsals. (For more information on breaching rehearsals, see *Chapter 4* of this manual or *FM 101-5*.) Time is allocated to each unit to perform combined-arms rehearsals. A rehearsal site should reflect the actual obstacle system in as much detail as possible. Friendly forces choose terrain similar to the operational area and construct a practice obstacle system based on OBSTINTEL. At a minimum, rehearsals should include a leader and key personnel walk-through and individual rehearsals by support, breach, and assault forces. As time permits, conduct as many full-scale rehearsals as possible. When possible, friendly forces rehearse the operation under the same conditions expected during the actual mission, including battle-space obscuration and darkness.

2-32. When a force commander rehearses a breaching operation, he also rehearses several contingency plans. The contingencies should include possible enemy counterattacks and attacks by enemy indirect-fire systems (artillery, rockets, attack helicopters, and other air assets). This also includes the enemy's use of NBC munitions. If updates become available after the last possible rehearsal, immediately pass the information to the affected elements.

EXECUTION

2-33. Each element of the breaching organization must accomplish its assigned tasks for a breaching operation to be successful. Additionally, the entire force must be flexible when responding to changes in the plan. Examples of changes are—

- Allocating additional assets to the support force due to attrition.
- Changing the location of the point of penetration.
- Modifying the scheme of reduction and changing the order of the units passing through the created lane.

2-34. All units must continually update the commander during mission execution. A critical phase of a breaching operation is the movement of the support force from the LD to its SBF positions. The support force must be prepared to maneuver to its SBF positions. Additionally, the support force must plan for the enemy's use of SCATMINEs along its axis of advance, which may require it to conduct a breaching operation.

2-35. After the support force has occupied its SBF positions and the commitment criteria of the breach force has been met (achieved necessary suppression and obscuration), the breaching commander should order the breach force to move forward to a specific point (6- or 8-digit grid) on the battle space to begin reduction. As the breach-force commander maneuvers his force to the point of breach, he must always remain cognizant of his relationship to the support force to ensure that he does not unnecessarily mask their fires.

2-36. Actions by the assault force at an assault position may include—

- Verifying current friendly and enemy situations using tactical reports from company teams or support forces.
- Issuing fragmentary orders (FRAGOs) and disseminating information to the lowest level.

- Confirming TRPs and direct-fire responsibilities.
- Positioning artillery observers.
- Conducting final prepare-to-fire checks.
- Reorganizing to compensate for combat losses.

2-37. The breach force commander should lead with his security element to allow it to establish nearside security before the reduction element arrives at the obstacle. As the security element occupies its position, the reduction-element commander should conduct a recon to acquire obstacle information and quickly confirm/refine his scheme of reduction. As the reduction-element commander brings his element forward, he should report the exact location where the obstacle will be reduced and the estimated time to reduce/proof and mark the obstacle. At the same time, the breach-force commander must assess the effectiveness of the suppression and determine if and how he will augment the support force's fires.

2-38. As the reduction continues, the support- and breach-force commanders must update their commander with the current status. Specifically, the support-force commander must update on his ammunition status and the strength of his unit. The breach-force commander must update on the progress of the reduction effort and the effectiveness of obscuration and suppression. The breach-force commander should verify who the assault force will be and where the assault force will come from so that he can assist in their passage.

2-39. As the reduction effort nears completion, the breach-force commander should report the grid coordinates of the far-recognition marker so that the assault force can begin movement to that location. When reduction is complete, the breach-force commander must—

- Report lane completion.
- Provide grid coordinates of the final-approach marker.
- Confirm/report the lane-marking pattern and material.

2-40. The breach-force commander establishes farside local security and assists in the assault force's passage. Once the lanes have been reported/signaled as being open for traffic, the assault force passes through the lanes to destroy forces that can bring direct fires on created lanes. To accomplish this mission, the assault force may have to reduce enemy protective obstacles.

2-41. The obstacle system continues to be a choke point and danger area even after the assault force has passed through created lanes. Additional lanes are constructed to speed the follow-on forces' passage, and lanes are monitored to ensure that they remain clear. Reduction assets need to be maintained near the points of breach in case a scatterable minefield is emplaced by the obstacle. The lane-marking pattern is upgraded to intermediate. There are many ways this can be accomplished. The higher HQ—

- Takes command of the point of breach and has some or all of the reduction element revert to its control and continue creating/expanding lanes in the obstacle for follow-on forces.

- Takes command of the point of breach and has additional reduction assets follow closely behind the assault force and create additional lanes to allow the reduction element to remain under unit control.
- Tasks the breaching unit to maintain lanes in tactical obstacles.

2-42. The lanes are widened to allow two-lane traffic through obstacles, and they are marked with the full lane-marking pattern. Deliberate marking and fencing systems are installed, and military police (MP) establish necessary traffic control. Eventually, follow-on engineer forces clear obstacles and eliminate the choke point.

Chapter 3

Lane and Bypass Marking

This chapter implements STANAG 2036.

This chapter provides commanders with an Army-wide standard system for lane and bypass marking. It centers around a systematic, phased upgrade of lane marking. Each upgrade conforms to tactical requirements for that phase of the attack, from initial obstacle reduction to the passage of larger follow-on forces and the return traffic necessary to sustain the force. Lane or bypass marking is a critical component of obstacle reduction. Effective lane marking allows commanders to project forces through an obstacle quickly, with combat power and C^2 intact. It gives an assault force and follow-on forces confidence in the safety of the lane and helps prevent unnecessary casualties.

LANE-MARKING SYSTEM

3-1. There are two critical components to any lane-marking system—the lane-marking pattern and the marking device.

- **Lane-marking pattern**. The location of markers indicating the entrance, the lane, and the exit.
- **Marking device**. The type of hardware emplaced to mark the entrance, the lane, and the exit.

3-2. The lane-marking system outlined in this chapter centers on standardizing the marking pattern across the Army rather than standardizing the marking device. Standardizing the marking pattern is critical to offensive operations. A common lane-marking pattern—

- Enables cross attachments and adjacent units to recognize lanes easily with minimal knowledge of a particular unit's tactical SOP.
- Gives all forces a standardized set of visual cues needed to pass through a lane safely and maintain their momentum.
- Facilitates quick conversion to the lane-marking requirements of STANAG 2036.

3-3. Until the Army adopts a standard lane-marking device, commanders decide what hardware to use. This gives units greater flexibility, allowing them to adopt a marking device that is tailor-made for their type of unit and mission (armored or light force, mounted or dismounted attack, limited visibility, thermal capability). Regardless of the device used, it must support the standard lane-marking pattern outlined in the following paragraphs. Therefore, commanders should consider the marking guidelines and examples in this chapter before developing or adopting their own system. Commanders should determine and standardize the marking device used based on the availability of resources and METT-TC.

LANE-MARKING TERMS

3-4. The following terms provide a common basis for discussing lane marking:

- Entrance markers.
- Handrail markers.
- Exit markers.
- Entrance-funnel markers.
- Final-approach markers.
- Far-recognition markers.
- TCPs and guides.

ENTRANCE MARKERS

3-5. Entrance markers indicate the start of a lane through an obstacle. They signify the friendly-side boundary of the obstacle and the point at which movement is restricted by the lane width and path. Entrance markers also indicate the lane width. Therefore, it is critical for a maneuvering force to distinguish the entrance point clearly, since it signifies the point at which passing vehicles can no longer adjust their movement in reaction to the situation (direct and indirect fires) without jeopardizing the force. Entrance markers must be visually different from handrail markers to help a force distinguish this critical point in the lane. The distance between entrance markers must be the same as the width of the created lane. Entrance markers placed a minimum of 4.5 meters apart indicate a lane capable of supporting mounted movement, while markers placed a minimum of 1 meter apart indicate a dismounted lane. When obstacle boundaries are vague or unknown, such as a buried minefield, the breach force uses its best judgment and marks the entrance where obstacle reduction and lane proofing begin.

HANDRAIL MARKERS

3-6. Handrail markers define the lane path through an obstacle and indicate the limits of the lane width. As a minimum, mounted and dismounted lanes have a left handrail. Mounted and dismounted forces moving through a lane should keep the left handrail immediately to the left of their vehicle or person. The lane width is defined by the entrance markers. Therefore, when only the left handrail is marked, drivers use entrance and handrail markers to gauge the lane width and path. As the operation progresses, lane marking may be upgraded to include left and right handrails.

EXIT MARKERS

3-7. Exit markers indicate the far side limit of a lane through an obstacle. For a passing force, the exit marker signifies the point at which movement is no longer confined to the lane path. Like entrance markers, exit markers must be distinguishably different from handrail markers; however, the exit may be marked the same as the entrance. Exit markers are placed to the left and right of an exit point and spaced the width of the reduced lane. This visual reference is critical when only a left handrail is marked. The combination of entrance, left-handrail, and exit markers gives the driver and the tank commander visual cues (entrance and exit points, lane width and path) to pass safely along a lane.

ENTRANCE-FUNNEL MARKERS

3-8. Entrance-funnel markers augment entrance marking. The V formed by the funnel markers forces approaching platoons into a column formation and assists drivers and tank commanders in making last-minute adjustments before entering the lane.

FINAL-APPROACH MARKERS

3-9. Final-approach markers are highly visible, more robust markers that augment the visual signature of entrance-funnel markers. Units must be able to see the entrance funnel and/or the entrance-funnel markers from the final-approach marker. The final-approach marker—

- Provides the assault-force commander with a highly visible reference point toward which to maneuver his formation.
- Signals company-team commanders to begin changing to a column formation.

FAR-RECOGNITION MARKERS

3-10. Far-recognition markers are highly visible markers located between the final-approach marker and the friendly unit. They are primarily used when passing battalion-size forces through a lane where direct observation of the final-approach marker is denied due to distance, visibility, or terrain. Far-recognition markers should be different from the final-approach marker. Far-recognition markers indicate the point at which forces begin changing their formation to posture for the passage. A single far-recognition marker may serve up to two lanes when located 200 to 400 meters apart. Once lanes are upgraded to two-way traffic, far-recognition markers are required for each two-way lane. Far-recognition markers should be visually alterable so that they can be distinguished from the far-recognition marker of an adjacent lane. This assists the C^2 of large formations when passing on several adjacent lanes. When a far-recognition marker serves more than one lane, a guide or a TCP is collocated with the far-recognition marker nearest the obstacle.

TRAFFIC-CONTROL POSTS AND GUIDES

3-11. A TCP or a guide is a two-man team with communications means that assists the commander in controlling the movement of forces. When available, consider using MPs to establish and man these posts. Initially, the commander may use other personnel as guides to man critical far-recognition markers until MPs establish full TCPs. TCPs and guides provide the commander with a means to control traffic flow to appropriate lanes. When there are multiple lanes branching off a single far-recognition marker, the TCP helps direct the formation to various lanes. The TCP can also assist in modifying the traffic flow when lanes have been closed for maintenance, for lane expansion, or by enemy SCATMINEs. The TCP or guide must give the commander the azimuth and distance to the final-approach marker, the device used for the approach marker, and the level of the lane-marking pattern. For light forces, guides may physically escort passing units from the far-recognition marker to the lane entrance. In short, the TCP allows the commander to make last-minute changes in the traffic flow, giving him increased flexibility to react to the enemy situation.

3-12. Another technique to facilitate the movement of the assault force to the points of breach is for the assault force to maneuver a recon element, such as a section, behind the breach force. This section observes the breach effort, identifies lanes, and leads the assault force to lanes after the breach is complete.

LEVELS OF LANE MARKING AND PATTERNS

3-13. There are three levels of lane marking—initial, intermediate, and full. Each lane-marking level provides an increase in lane signature and capability. Lane requirements change as a breaching operation matures from the passage of the assault force to the passage of larger follow-on forces. Initial lane-marking requirements are driven by the nature of the fight through the obstacle. Marking must be rapid, providing only the bare minimum signature needed to safely pass small units (company teams, platoons) that make up the assault force. This contrasts with lane requirements during later phases of an attack where larger units (battalion and above) are passed to subsequent objectives and mark improvements to be made. Two-way traffic becomes a priority for the simultaneous forward passage of combat units and return traffic (ambulances, empty supply vehicles) necessary to sustain the force. With the increase in traffic volume comes more diverse forces and levels of driver experience. Lane-marking limits must be clear to the most inexperienced driver or crewman; do not assume he has a knowledge of the unit's SOP. A fully developed lane must support two-way traffic and be completely marked.

3-14. Commanders must be aware of how the lane needs of a force change with the operation so that they can anticipate lane-marking and lane-capability requirements. Integrating the levels of lane marking into the overall breaching plan ensures that a unit's needs are satisfied. The forces necessary to mark and upgrade lanes must be allocated and tasked with that mission. The scheme of maneuver and the service-support plan are the basis for analyzing lane requirements. The following paragraphs describe lane-marking patterns in detail and provide guidelines on when a commander should upgrade lane marking and lane capability.

INITIAL LANE MARKING

3-15. The breach force emplaces the initial lane-marking pattern immediately after the obstacle is reduced and, if required, the lane is proofed. This is a signal to the assault-force commander that the lane is ready for traffic. Initial lane marking is kept to a minimum, focusing on the markings needed to pass immediate assault forces (company teams and smaller) through the lane to seize the point of penetration on the farside objective. Normally, the assault force can observe the lane as it is being marked and does not need the greater visual signature of mature lane marking. The initial lane-marking pattern contains—

- Entrance markers.
- Left-handrail markers.
- Exit markers.
- Entrance-funnel markers.
- A final-approach marker.

3-16. *Figure 3-1*, page 3-6, shows the initial lane-marking pattern. Entrance, left-handrail, and exit markers are the first markers that the breach force emplaces since these markers define the location and limits of the lane. These markers should be emplaced as follows:

- Entrance markers are placed to the left and right of the reduced lane's entrance point. They are spaced the width of the lane (at least 4.5 meters for mounted lanes and 1 meter for dismounted lanes).
- Left-handrail markers are placed at the left limit of the lane along the entire path. They are placed at 15-meter intervals for mounted forces and 5-meter intervals for dismounted forces. Commanders can modify the intervals based on the terrain, the visibility, the lane length, and the lane path. Additionally, commanders may choose to mark left and right handrails when the lane path is lengthy or unclear (through a complex obstacle).
- Exit markers are placed to the left and right of the lane at the far side of the obstacle (at least 4.5 meters for mounted lanes and 1 meter for dismounted lanes).

3-17. After entrance, left-handrail, and exit markers are placed, the breach force emplaces entrance-funnel markers and the final-approach marker. These markers enhance the visual signature of the lane entrance for the assault force. If the assault force is in a position to observe the obstacle reduction, the commander does not have to wait until entrance-funnel markers and the final-approach marker are set to commit the assault force. However, the breach force continues to install the necessary markers to improve the lane signature for larger or subsequent follow-on forces maneuvering toward the lane. Entrance-funnel markers and the final-approach marker are emplaced as follows:

- Entrance-funnel markers are placed at 15-meter intervals for mounted forces and 5-meter intervals for dismounted forces. They are placed diagonal to the lane entrance to form a 45-degree V as shown in *Figure 3-1*.
- The final-approach marker is centered on the lane and placed at least 200 meters from the lane entrance for mounted forces. For dismounted forces, the nature of the attack may preclude using a final-approach marker initially. Place a final-approach marker 30 meters from the entrance as soon as the mission allows. Commanders modify the recommended distance for the final-approach marker based on the terrain and the visibility.

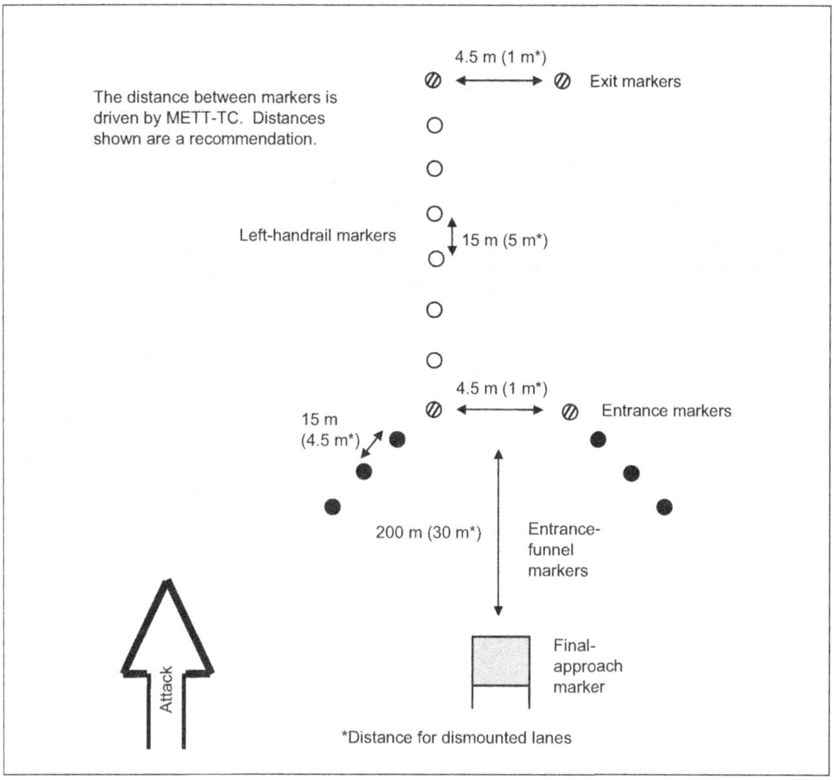

Figure 3-1. Initial Lane-Marking Pattern

INTERMEDIATE LANE MARKING

3-18. Upgrading initial lane marking to intermediate lane marking is triggered by one of two key events—

- Commitment of larger combat forces who are unable to directly observe the obstacle as it is reduced.
- Rearward passage of sustainment traffic (casualty evacuation and vehicle recovery).

3-19. Intermediate lane marking has two goals—

- Increasing the lane signature to assist in the passage of larger combat forces.
- Providing sufficient marking for two-way, single-lane traffic.

3-20. Intermediate lane marking builds on the initial lane-marking pattern by adding right-handrail markers, exit-funnel markers, far-recognition markers, and a farside final-approach marker. *Figure 3-2* shows the intermediate lane-marking pattern.

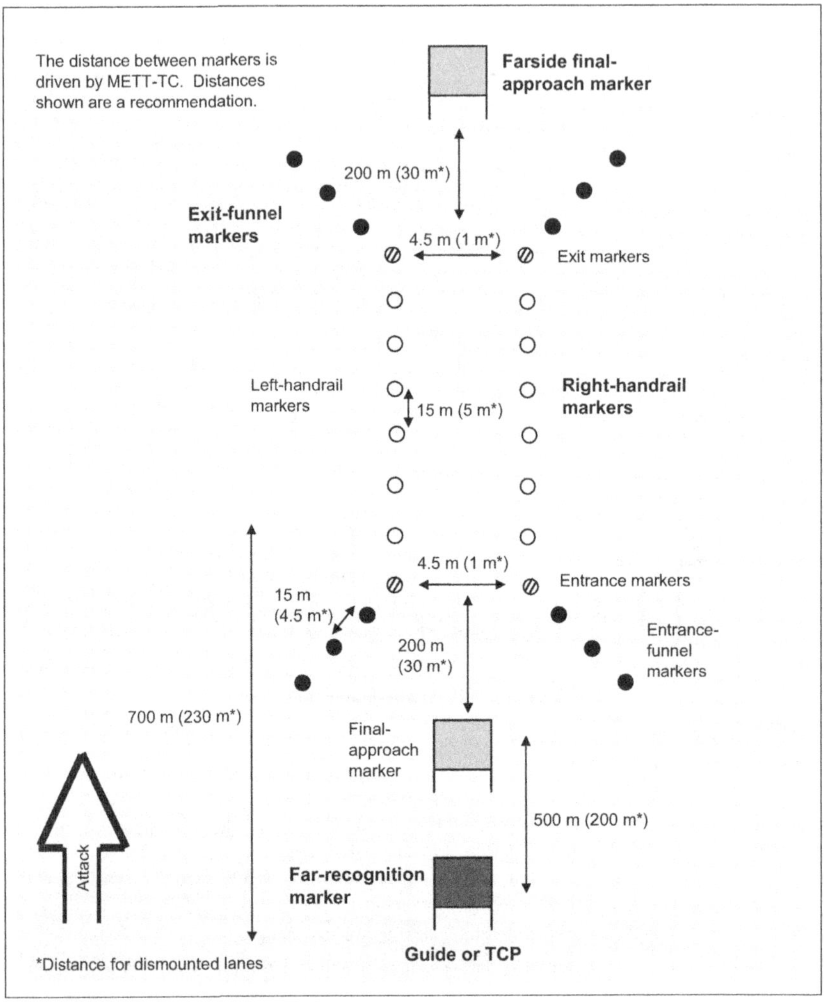

Figure 3-2. Intermediate Lane-Marking Pattern

3-21. The commander sets the priority of marker emplacement based on the situation. If the scheme of maneuver requires the immediate passage of larger combat forces, right-handrail markers and the far-recognition marker may be the priority. On the other hand, if it is necessary to ground-evacuate casualties or to recover vehicles, it may require emplacing right-handrail markers, exit-funnel markers, and the farside final-approach marker first. Intermediate lane marking gives the commander greater flexibility. Establishing a network of guides becomes increasingly important to direct the flow of forward and returning forces.

3-22. Regardless of the reason for an upgrade, the first step is to emplace right-handrail markers. These markers define the rightmost limit of the lane. They are placed the width of the lane as defined by entrance and exit markers. The right handrail follows a path parallel to the left handrail through the obstacle. Right-handrail markers are placed at the same interval as left-handrail markers.

3-23. Exit-funnel markers and a farside final-approach marker are emplaced to mirror entrance markers. For the forward passage of large combat forces, exit-funnel markers prevent the premature deployment of a passing force into combat formation before it is safely outside the obstacle. Exit-funnel markers also become entrance-funnel markers for rearward passing traffic, giving these forces the visual cues needed to align themselves on the lane. Exit-funnel markers are augmented by a farside final-approach marker to help rearward passing forces clearly identify the lane from their side. The farside final-approach marker is centered on the lane and placed 200 meters (30 meters for dismounted forces) from the exit (farside entrance).

3-24. A far-recognition marker completes the intermediate lane-marking pattern. It provides commanders with a visual signature or a series of signatures on which to guide their movement toward the lane. When the assault force is moving over a greater distance to the lanes, additional far-recognition markers may be required. For mounted forces, the far-recognition marker nearest the obstacle is placed a maximum of 500 meters from the final-approach marker (the final-approach marker must be visible from the far-recognition marker). Dismounted forces may require a system of guides instead of far-recognition markers for passing combat forces. However, far-recognition markers must be emplaced as soon as possible to reduce guide requirements for passing mounted sustainment traffic. This gives the assault-force commander the space needed to transition to a column formation.

3-25. A commander collocates guides or TCPs at the far-recognition marker when he feels the situation requires more positive control over traffic flow. Normally, the need to increase traffic control comes at the same time as the need for limited two-way traffic and intermediate lane marking. Commanders should plan for the use of dedicated guides once they have upgraded to intermediate-level marking. Guides or TCPs become mission critical during limited visibility or in restrictive terrain. They should also be used when a single far-recognition marker feeds more than one lane. TCPs must be manned with a minimum of two soldiers and must have secure frequency-modulated (FM) communications with the controlling HQ. It is essential that soldiers acting as guides or TCPs know the—

- Azimuth and distance to the final-approach marker or the 8-digit grid coordinate of the final-approach marker that is entered into the unit's Global Positioning System (GPS) receiver.
- Lane-marking pattern currently emplaced.
- Type of final-approach marker used.
- Traffic-control plan and march order.

Guides must be kept up-to-date on the status of lane marking, maintenance, and closures.

FULL (TWO-WAY) LANE MARKING

3-26. Lane marking is upgraded to a full lane-marking pattern when the commander desires uninterrupted two-way traffic flow. Expanding lanes to full lane marking is resource-intensive and is not part of a breaching operation. A full lane supports uninterrupted two-way traffic. The marking pattern is modified to give forces passing forward or rearward the same visual signature.

3-27. Upgrading to a full lane-marking pattern is normally assigned to follow-on engineer forces, since it is usually beyond the immediate capability of engineers with forward units. In special cases, the commander may task engineers with forward units to expand and upgrade a lane from an intermediate to a full lane-marking pattern. However, the trade-off is decreased support to forward units and delays associated with changing task organization. The full lane-marking pattern is also used when marking a lane through friendly obstacles along a main supply route (MSR) or a passage lane.

3-28. Upgrading intermediate lane marking to full lane marking begins by temporarily closing the lane, rerouting traffic, and expanding the lane width. The lane is always expanded to the left in relation to the direction of the attack. Engineers reduce and proof the obstacle beginning at the left handrail to give a total lane width of 10 meters (5 meters each way). The expansion width requirement is the same for armored and light forces, since both forces must be able to pass mounted sustainment and combat forces. When the lane is expanded by mechanical means, it is necessary to entirely remove the left-handrail markers during expansion. Whenever possible, these markers should be left alone to serve as a reference point during lane expansion. If removed altogether, left-handrail markers must be replaced since they form the left handrail for both forward and rearward passing forces.

3-29. When the lane width has been expanded to 10 meters, entrance, exit, right-handrail, funnel, and final-approach markers are replaced on the return lane. All markings are the same as described in previous paragraphs. *Figure 3-3*, page 3-10, shows a full lane-marking pattern.

3-30. The full lane-marking pattern has three entrance and three exit markers. Entrance and exit markers are placed the width of forward and return lanes and are visually different from other markers. Three entrance markers indicate a two-way traffic lane, and units should always use the rightmost lane. Entrance- and exit-funnel markers are placed slightly different from previous marking patterns. In the full lane-marking pattern, funnel markers extend out from the entrance and exit markers on the outside

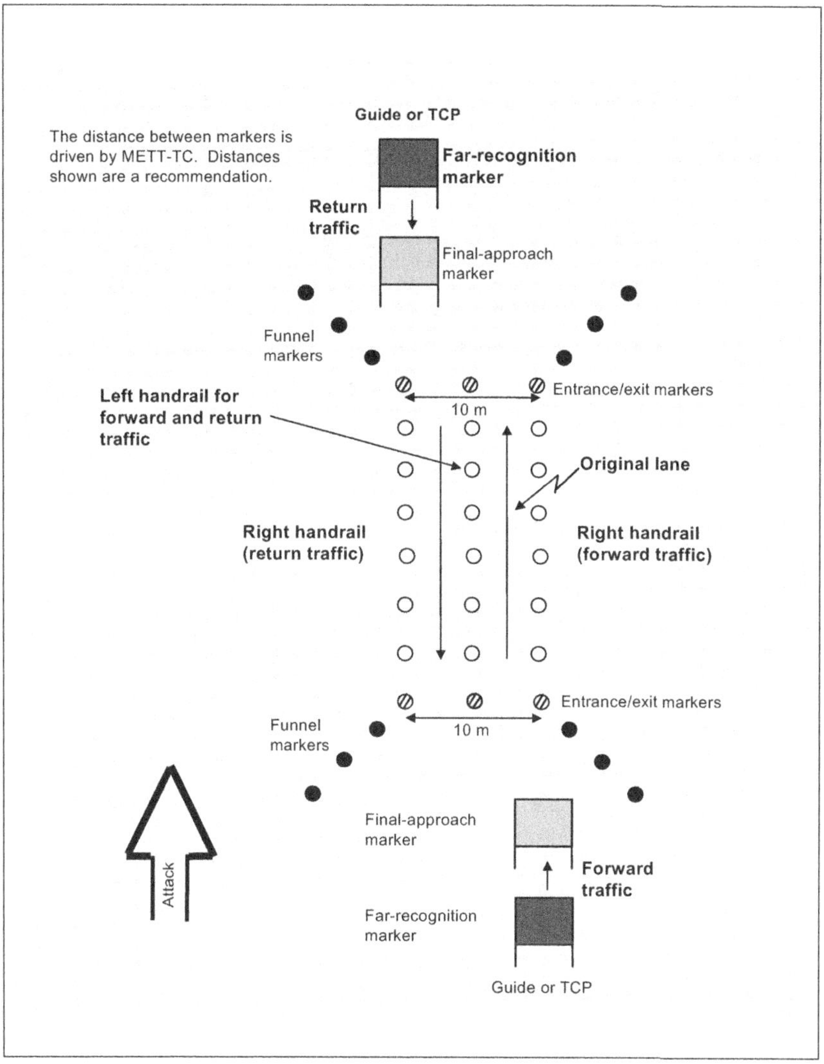

Figure 3-3. Full Lane-Marking Pattern

only. Final-approach markers are placed 200 meters from and centered on entrances of forward and return lanes. This helps forces identify entrance points from either direction. Far-recognition markers are placed a maximum of 500 meters from forward and return final-approach markers. Again, distances may be modified based on the terrain, the visibility, and the mission.

COMMANDER'S GUIDANCE FOR LANE MARKING

3-31. *Table 3-1* provides the commander with a summary of lane-marking levels, guidelines on unit responsibilities, and events that trigger a lane-marking upgrade. The chart focuses on TF-level breaching operations. In the table, *who* refers to the unit responsible for lane marking, *when* describes events that trigger the need to upgrade, and *markers* highlight a quick reference for lane-marking requirements.

Table 3-1. Lane-Marking Levels, Unit Responsibilities, and Trigger Events

	Initial	Intermediate	Full (Two Way)
Who	TF breach force	TF breach force	Brigade
When	Obstacle is reduced	Passing battalion- or company-size forces	Passing brigade- or battalion-size forces
	Passing platoon- or company-size assault forces	Passing force which cannot see the lane	Situation requires uninterrupted sustainment traffic
		Passing TF combat trains	
Markers	Entrance	Add right handrail	Expand lane width to 10 meters
	Exit	Add exit funnel	Adjust entrance/exit
	Left handrail	Add farside final approach	Adjust left/right handrails to new width
	Entrance funnel	Add far recognition	Add far recognition
	Final approach	Add guides or TCPs	Add farside guides or TCPs

MOVEMENT THROUGH THE LANE

3-32. Standard lane-marking patterns help the commander in two critical aspects when moving a unit through a lane:

- Lane markers help the force posture its formation for efficient passage through the defile that is caused by the lane through the obstacle.
- The combination of lane markers and TCPs gives the commander greater flexibility in the forward and rearward movement of forces.

3-33. Far-recognition and final-approach markers in a standard lane-marking pattern help commanders smoothly change assault and follow-on forces into a column formation before passing through the lane. In most cases, the situation should allow the unit to move in a column formation from the outset; however, the enemy situation and the terrain may demand that forces remain dispersed as long as possible. Far-recognition, final-approach, and entrance-funnel markers are used to trigger commanders to alter their formations.

Figure 3-4 shows how formations at various levels may change to efficiently pass through lanes. A battalion TF initially moves from its assault position in a box, V, or wedge formation depending on the number of lanes being used.

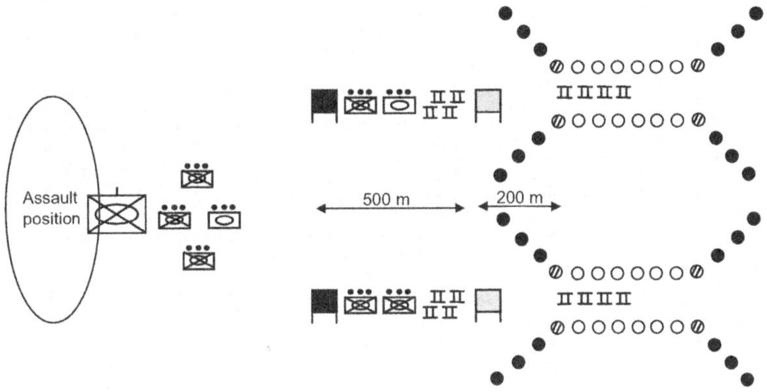

	Move out of assault position	Visually identify far-recognition marker	Pass far-recognition marker	Pass final-approach marker	Pass through exit-funnel markers
Battalion	Box (2 lanes) V (3 lanes) Wedge (3 lanes)	Box (2 lanes) V (3 lanes) Wedge (3 lanes)			
Company	Any combat formation	Column with dispersal	Column	Column	Resume formation
Platoon	Any combat formation	Any combat formation	Column with dispersal	Column	

Figure 3-4. Formation Changes Based on Lane Markers

3-34. Initially, company teams and below may move in any combat formation. Once lead elements of the battalion visually identify the far-recognition marker, company teams begin moving into a column formation with platoons still deployed. Platoons—

- Change into a column formation with dispersal as they move from the far-recognition marker to the final-approach marker.
- Change into a column formation as they move from the final-approach marker to entrance markers, using the entrance-funnel markers as guides.
- Redeploy once they pass through the exit or the exit funnels.

3-35. As the operation progresses, commanders should upgrade lane marking and establish traffic control as early as METT-TC allows. The combination of upgrading lanes from the initial marking pattern and establishing TCPs gives the commander greater flexibility. Normally, TCPs are established at forward and return far-recognition markers. Upgrading to full lane marking that is capable of supporting uninterrupted two-way traffic is generally determined by the need to pass increased volumes of traffic.

3-36. *Figure 3-5*, page 3-14, shows the flexibility that full lane marking and TCPs provide. In this example, friendly forces have created two lanes, 200 meters apart. Lane 1 is a full lane, and lane 2 is an intermediate lane. A far-recognition marker on the nearside and the farside can feed traffic to both lanes. Full lane marking and TCPs allow the commander to continue pushing combat power forward on lanes 1 and 2 simultaneously. At the same time, the farside TCP directs all return traffic to lane 1. Commanders should ensure that recovery vehicles are on site to remove immobile and destroyed vehicles.

LANE MAINTENANCE AND RESPONSIBILITIES OF TRAFFIC THROUGH LANES

3-37. As discussed in *Chapter 2*, units must plan for lane maintenance (widen lanes, upgrade lane markings, ensure that lanes remain open, ensure C^2 of traffic through the breach area). The responsibility for lane maintenance should be delineated in unit orders to ensure alignment with the scheme of maneuver. Specifically, units must determine who is responsible for—

- Maintaining lanes in tactical and protective obstacles.
- Maintaining lanes when a change occurs.
- Controlling traffic within the breach area.

3-38. *Figures 3-6 through 3-9*, pages 3-15 and 3-16, show how a unit defines responsibilities for lane maintenance and traffic control. In this example, a TF breaching operation is being conducted. Teams A and B provide suppressive fires as the support force, Team C reduces tactical obstacles as the breach force, and Team D seizes OBJs LION and TIGER as the assault force. The brigade intends to pass an armored TF through obstacles as the first element of the follow-on forces once the TF has cleared a route through the breach area. Note that the initial-lane requirement to pass Team D (the assault force) is one vehicular lane, but the breach force breaches a second lane to pass follow-on forces (an armored TF).

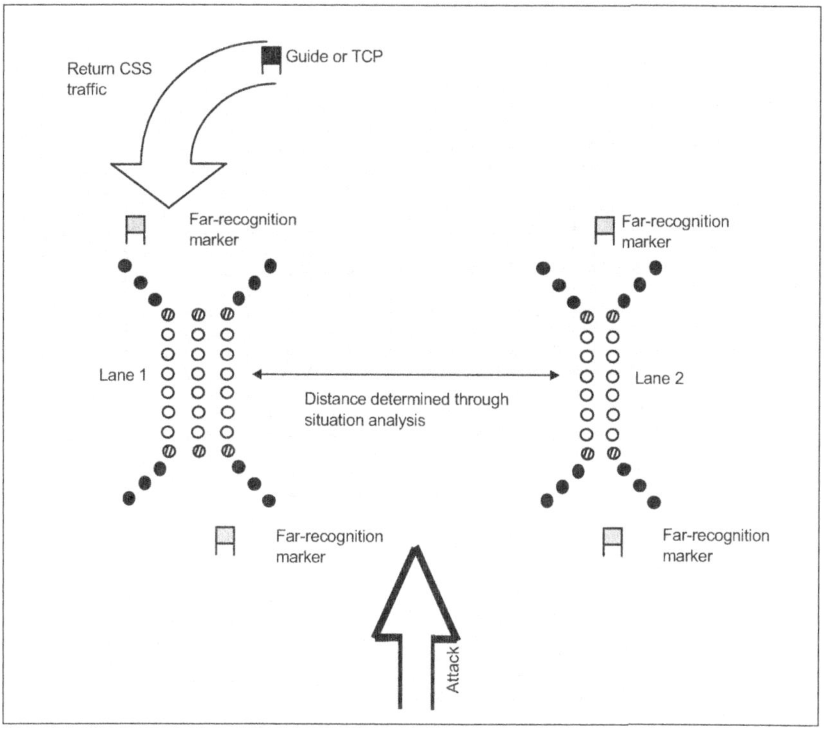

Figure 3-5. Multiple Lanes Used at Different Levels

3-39. As the breaching operation progresses, the breach force creates two lanes in the tactical obstacles. The breach force marks lanes with the initial lane-marking pattern and passes the assault force through the lanes (*Figure 3-6*). At this stage of the breaching operation, the TF maintains lanes in the tactical obstacles and controls all movement of forces within the breach area.

3-40. The assault force passes through the tactical obstacles and attacks to seize OBJs LION and TIGER. As the assault force attacks, it encounters protective obstacles and reduces and marks them. Immediately after the assault force passes through the tactical obstacles, the brigade assumes responsibility for lane maintenance in the tactical obstacles. The brigade also assumes responsibility of traffic control from the beginning of the breach area through the tactical obstacles. This includes creating additional lanes, ensuring that the lanes remain open, upgrading lane markings at the tactical obstacles, establishing TCPs, and calling units forward into the breach area. The TF is responsible for maintaining lanes through the protective obstacles (*Figure 3-7*).

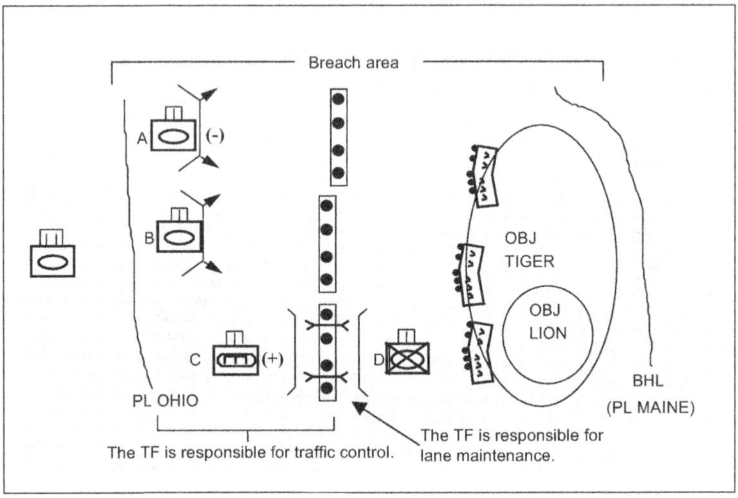

Figure 3-6. Assault Force Passes Through Tactical Obstacles

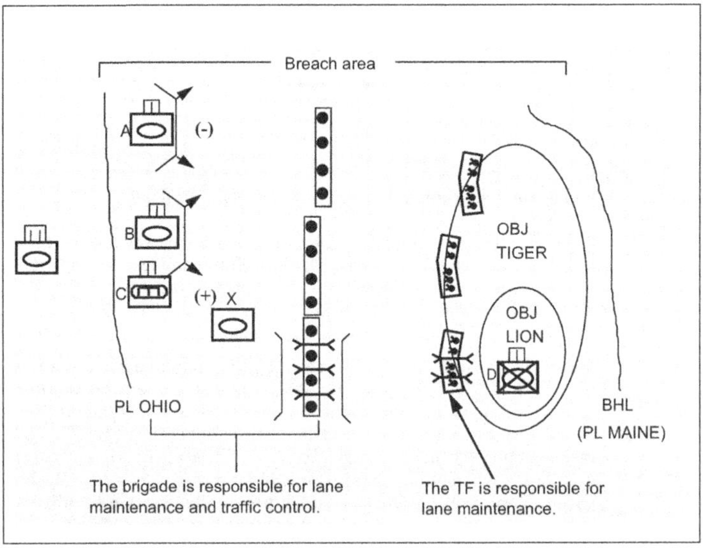

Figure 3-7. Brigade Assumes Control of Lanes in Tactical Obstacles

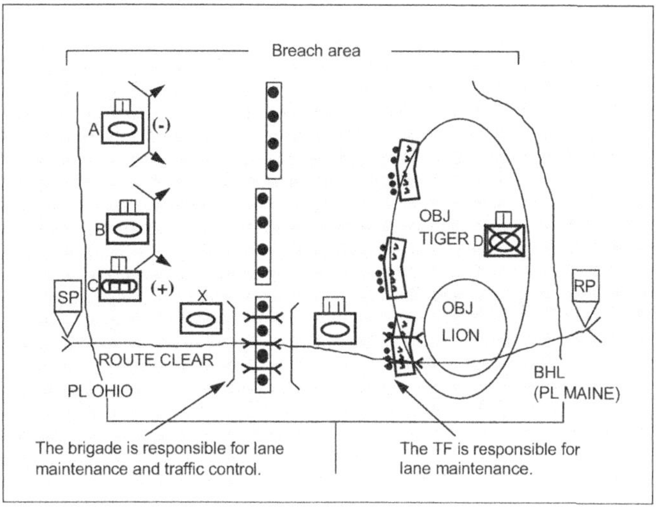

Figure 3-8. Follow-On Forces Begin to Pass Through the Breach Area

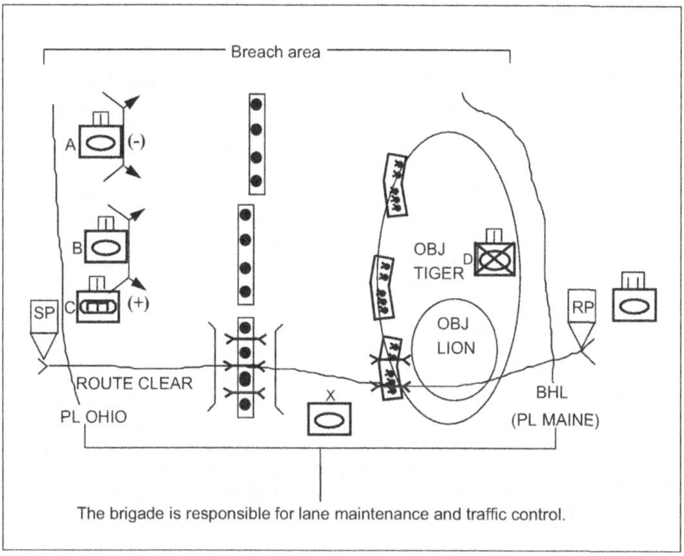

Figure 3-9. Brigade Assumes Full Control of the Breach Area

3-41. Once the assault force has seized its objective and reported that it has established a clear route up to the BHL (PL MAINE), the brigade begins to pass the follow-on TF through the breach area. At this point, the brigade controls traffic throughout the breach area. The TF is responsible for maintaining lanes in the protective obstacles (*Figure 3-8*).

3-42. As the follow-on forces cross the BHL, the breaching operation is complete; however, the obstacles within the breach area remain a constriction for other follow-on forces. At this point, the brigade maintains lanes and controls traffic. The brigade begins to focus on establishing MSRs through the breach area using corps engineer assets, which are task-organized to the brigade from the division (*Figure 3-9*).

LANE-MARKING DEVICES

3-43. Currently, the Army has no standard lane-marking device. The current marking systems in the Army's supply system are—

- Hand-Emplaced Minefield Marking System (HEMMS), national stock number (NSN) 9905-01-019-0140.
- Minefield Marking Set Number 2, NSN 9905-00-375-9180.
- Cleared-Lane Marking System (CLAMS), NSN 2590-01-205-3082.

3-44. Most lane marking in the field is done with nonstandard marking devices. Before adopting a nonstandard marking device, commanders should consider the guidelines summarized in *Table 3-2*. Coordination must occur between forces who are tasked to mark lanes and forces who are expected to use lanes to ensure that the marking material will work. For example, if a light force plans to create and mark vehicle lanes with an infrared light, vehicle drivers with enhanced (thermal) night-vision devices may not be able to see the markings.

Table 3-2. Guidelines for Lane-Marking Devices

Markers	Mounted Forces	Dismounted Forces
Handrail and funnel	Visible by TC and driver (buttoned up) from 50 meters	Visible by dismounted soldier in prone position from 15 meters
	Quick and easy to emplace (minimizes need to expose soldiers outside the emplacement vehicle)	Lightweight, quick and easy to emplace (dismounted soldier should be able to carry enough markers for the lane and still be able to fire and maneuver)
Entrance and exit	Visible by TC (buttoned up) from 100 meters	Visible by dismounted soldier from 50 meters
	Visually different from handrail and funnel markers	Visually different from handrail and funnel markers
	Quick and easy to emplace (may require soldier to dismount to emplace)	Lightweight, quick and easy to emplace
	Easily man-portable	
Final-approach and far-recognition	Visible by TC (not buttoned up) from 500 meters	Visible by dismounted soldier (on march) from 100 meters
	Visually different from each other	Visually different from each other
	Visually alterable to facilitate traffic control through multiple lanes	Visually alterable to facilitate traffic control through multiple lanes

3-45. *Figure 3-10* shows some of the devices that units currently use; they are easily procured or fabricated. This is by no means a complete listing; rather, it is intended to show commanders some of the options. Some of the general requirements for lane marking are—

- Markers must be able to withstand the rigors of the terrain, the weather, and the battle space.
- Markers should be easy to modify when visibility is limited, with minimal addition of manpower and equipment.
- Enhancements for limited visibility should be a constant source rather than a pulsating strobe. Strobes do not make the marking pattern readily apparent, particularly when approaching from an angle.

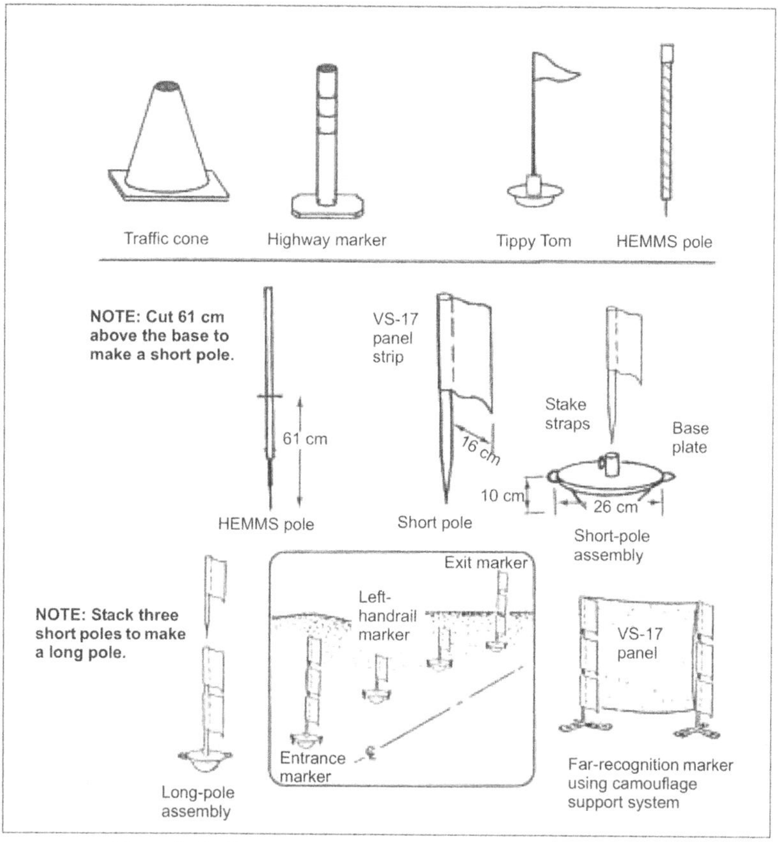

Figure 3-10. Examples of Nonstandard Marking Devices

BYPASS MARKING

3-46. Bypasses are marked using the same lane-marking patterns, marking devices, and visual cues as a lane. However, marking a bypass around an obstacle is much different from marking a lane through an obstacle because the amount of room available for forces to maneuver must be determined. While lanes through an obstacle are normally expanded to the left, bypasses are expanded away from the obstacle.

3-47. Bypasses must show where it is safe to rapidly pass around an obstacle, without unnecessarily restricting the force's ability to maneuver (*Figure 3-11,* page 3-20). To accomplish this, the marking unit will mark the bypass with a bypass-marking pattern (modified initial lane-marking pattern with right-handrail markers added). The unit will recon and initially mark the bypass 500 meters wide to allow platoons to maneuver around the obstacle. The width of the bypass is increased as time allows further recon. Limited visibility or additional obstacles discovered at the flank of the obstacle that is being bypassed may constrain the marked width of the path to less than 500 meters. (The right-handrail markers must be visible from the left-handrail markers.) In addition to bypass marking, units should consider establishing mobile guards on the friendly side of the obstacle to prevent vehicles/ personnel from entering it. All units on the battle space should possess marking material and understand how to emplace the bypass-marking pattern.

MARKING REQUIREMENTS UNDER THE NORTH ATLANTIC TREATY ORGANIZATION

3-48. The following paragraphs paraphrase the breach lane-marking requirements outlined in North Atlantic Treaty Organization (NATO) STANAG 2036. They also establish procedures used by US forces to modify intermediate and full lane-marking patterns to STANAG standard. The trend in modern warfare is increasing toward combined operations, so commanders need to be aware of their responsibilities under the NATO standard for marking hazardous areas, particularly lanes through obstacles.

COMMANDERS' RESPONSIBILITIES

3-49. STANAG 2036 outlines the minimum requirements for the lane-marking pattern before troops of other nations use the lane. Commanders must plan to convert the lane to the NATO marking standard as early as possible. When converting to the NATO standard, STANAG 2036 directs commanders to use the lane-marking devices as stated below. Within an offensive operation, this does not normally occur until the lane is matured to a full lane.

MARKING PATTERNS AND DEVICES

3-50. The intermediate lane-marking pattern satisfies the minimum lane-marking pattern that must be used before forces from another country pass through a lane. STANAG 2036 states that the entrance point, the exit point, and left and right handrails are the minimum required lane signatures.

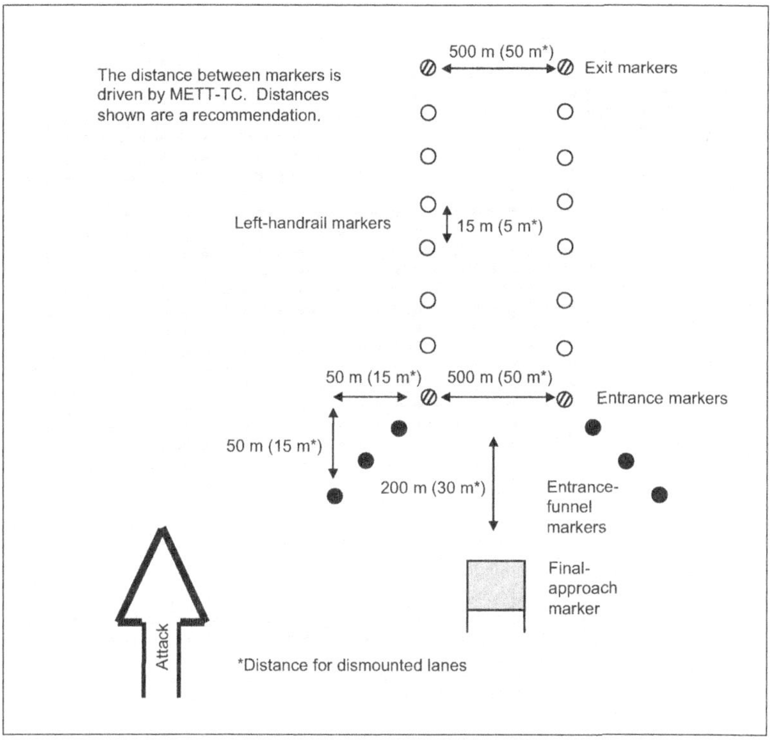

Figure 3-11. Bypass-Marking Pattern

Furthermore, STANAG 2036 requires that the entrance and exit of a lane be distinguishably different from other markers and that handrail markers be placed at intervals no greater than 30 meters. These requirements are met once lane marking is upgraded to the intermediate or full lane-marking pattern. Once the lane is marked to the the intermediate level, commanders can use the lane for allied forces to pass through without any additional marking if proper marking material is used.

3-51. *Figure 3-12* shows a NATO standard marker. The marker is placed at a right angle to the direction of travel so that the white portion of the arrow points inward to the lane, indicating the safe side on which to pass. The red portion points outward, indicating the lane limit or dangerous side of the lane. Units may fabricate NATO markers if they cannot get them through the supply system. STANAG 2036 requires that markers be visible from 50 meters under all conditions.

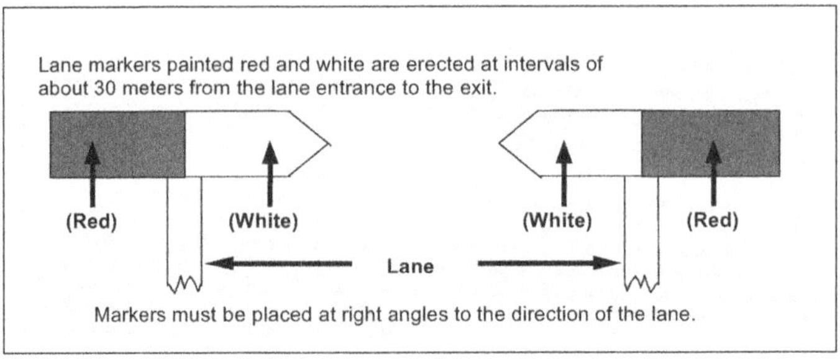

Figure 3-12. NATO Standard Marker

3-52. To convert intermediate lane marking to the NATO standard, affix NATO markers to long pickets and replace the existing entrance, exit, funnel, and handrail markers one for one. *Figure 3-13*, page 3-22, shows how NATO markers are used to convert an existing intermediate lane-marking pattern. Two NATO markers are used for entrance and exit markers to make them distinctly different. One NATO marker is affixed to each funnel marker and to each left- and right-handrail marker.

3-53. NATO uses white or green lights to illuminate markers at night. Entrance and exit markers are marked with two green lights or two white lights, placed horizontally, so that safe and dangerous markings on them are clearly visible. One white light or one green light is used on funnel and handrail markers. Commanders decide whether the light is placed on top of the NATO marker or placed so that it illuminates the marker (*Figure 3-14*, page 3-23). Lights must be visible from 50 meters under all conditions. Light sources for nighttime markers must have a continuous life of 12 hours.

COMPLEX OBSTACLES

3-54. When marking lanes through a complex obstacle, units can mark each obstacle independently or mark the entire obstacle system at once. The depth of the obstacle system, the distance between the obstacles, and the amount of marking material available determine which method is employed. See *Figures 3-15* and *3-16*, page 3-23, for examples of how complex obstacles may be marked; the examples use the initial lane-marking pattern.

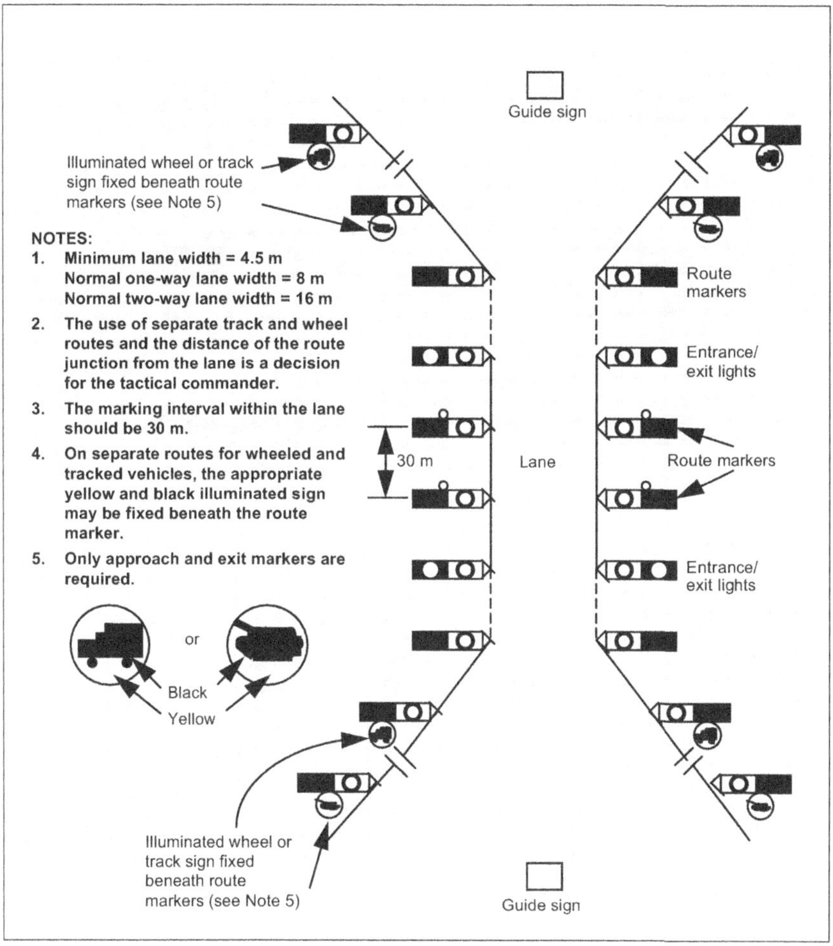

Figure 3-13. NATO Lane-Marking Standard

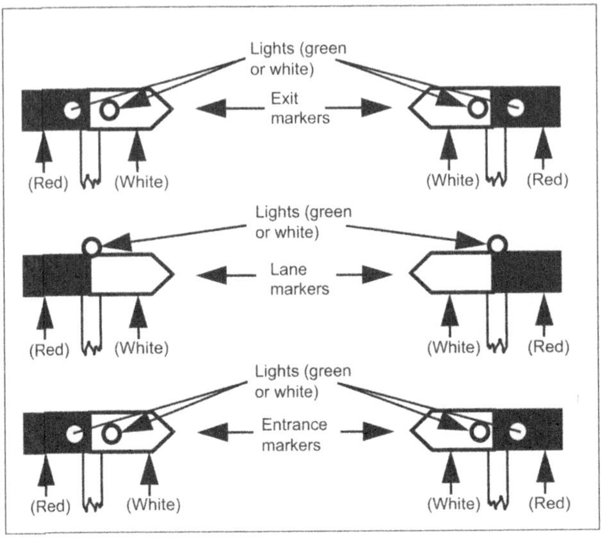

Figure 3-14. NATO Standard for Marking Lanes at Night

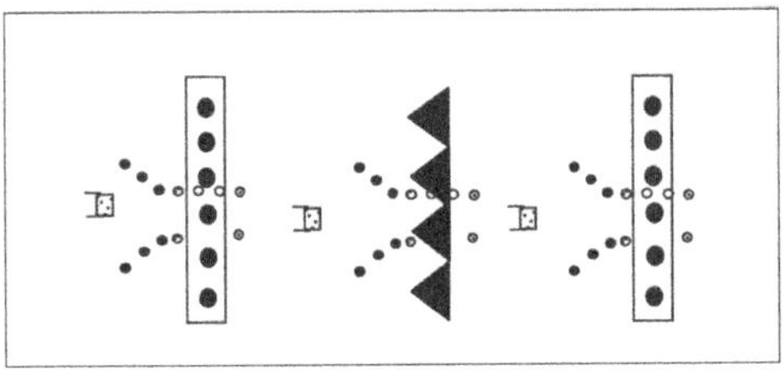

Figure 3-15. Marking a Complex Obstacle (Each Obstacle Independently Marked)

Figure 3-16. Marking a Complex Obstacle System

Chapter 4

Breaching Rehearsals

FM 101-5 gives a complete discussion of rehearsals, including types, techniques, and responsibilities. This chapter specifically discusses breaching rehearsals and considerations when conducting breaching rehearsals.

PLANNING

4-1. Rehearsals do not just happen; they must be well-planned. Successful rehearsals minimize the cost of time and maximize the understanding of the plan. Units may use a WO to establish responsibilities for preparing the rehearsal site. The WO can also identify battle drills that subordinate units should rehearse to prepare for the upcoming breaching operation. Units should include a complete rehearsal plan with an operation order (OPORD). The rehearsal plan may include a separate set of graphics if necessary.

TYPES

4-2. The five types of rehearsals are—

- Confirmation brief.
- Back brief.
- Combined-arms rehearsal.
- Support rehearsal.
- Battle drill or SOP rehearsal.

The combined-arms rehearsal and the battle drill or SOP rehearsal are most commonly used for breaching operations.

COMBINED-ARMS REHEARSAL

4-3. The maneuver unit HQ normally conducts the combined-arms rehearsal, and it is performed after subordinate units have issued their OPORD. The rehearsal ensures that subordinate plans are synchronized with those of other units and that the plans achieve the intent of the higher commander. A combined-arms rehearsal is particularly important when preparing for a complex breaching operation. It is very difficult to synchronize the actions of all the units involved in the operation.

4-4. Units should rehearse each phase of the operation, from the movement to and occupation of SBF positions through the passage of follow-on forces. The rehearsal must ensure that each unit knows how its tasks are integrated into the overall plan, in time and in space. For example, the support force must understand how long the breach force expects the reduction effort to take, so that the support force can ensure that its plan to suppress the enemy is adequate. At the same time, the breach force must understand its relative position to the support force when it is at the obstacle location. This ensures that the breach force, while moving to and reducing the obstacle, does not unnecessarily mask the fires of the support force.

4-5. At a minimum, the following should be addressed during a combined-arms rehearsal:

- The latest information concerning the enemy obstacle system.
- Actions that the higher HQ is taking to assist the breaching operation.
- The time available for emplacing smoke and activating CFZs (include location and duration).
- The criteria for lifting/shifting direct and indirect fires.
- The commitment criteria for the breach force.
- The method used to reduce the obstacle system.
- The marking to be used. One technique is to have a marked lane near the rehearsal site. All participants should view and drive/walk through the marked lane when arriving at or departing from the rehearsal site. The marking system should also be shown at support rehearsals conducted by combat service support (CSS) elements.
- The criteria and signals for the assault force to begin moving to created lanes.

BATTLE DRILL

4-6. A battle drill or an SOP rehearsal ensures that all participants understand techniques and procedures. These rehearsals are used by all echelons; but they are used most extensively by platoons, squads, and sections. For breaching operations, a battle drill or an SOP rehearsal can include lane-marking procedures or dismounted reduction techniques. Rehearsals should be conducted often to reinforce procedures that will be used during the breaching operation.

TECHNIQUES

4-7. Rehearsal techniques are limited only by the unit's resourcefulness. Generally, six techniques are used—

- Full dress.
- Reduced force.
- Terrain model.
- Sketch map.
- Terrain map.
- Radio.

Time and resource requirements for these techniques range from extensive to minimal. As the techniques are listed above, each successive one takes a decreasing amount of time and resources to prepare and conduct. Each rehearsal technique provides a different degree of understanding to participants. *Figure 4-1* shows the rehearsal techniques relative to time, resourcing, participation, and understanding.

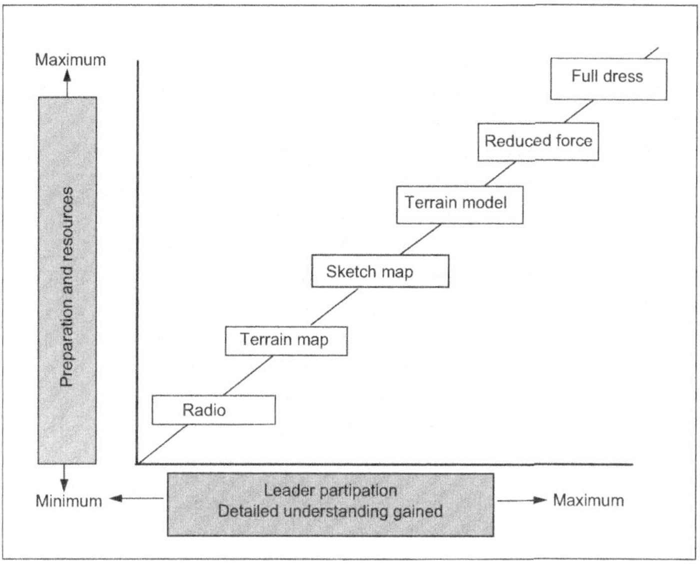

Figure 4-1. Rehearsal Techniques

4-8. When preparing for a breaching operation, use rehearsal techniques that provide the best understanding. Therefore, conduct a full-dress rehearsal if time and resources permit. Regardless of the rehearsal technique used, commanders must ensure that conditions (enemy situation, obstacle composition, visibility) are replicated in a realistic manner. For example, if a breach force is conducting a full-dress rehearsal, it should use the same obstacle system in the rehearsal as the one intelligence indicates that the enemy has emplaced. If intelligence is unavailable, the breach force should use the templated obstacle system. This requires the breach force to apply the same reduction techniques in the rehearsal that will be used during the operation.

4-9. Additionally, it is imperative that high standards be used during rehearsals. The breach force should rehearse proofing and marking obstacles. The assault force must rehearse all aspects of its assault, including the dismounted, ground fight if it is expected during the operation.

Appendix A

Metric Conversion Chart

This appendix complies with current Army directives which state that the metric system will be incorporated into all new publications. *Table A-1* is a conversion chart.

Table A-1. Metric Conversion Chart

US Units	Multiplied By	Metric Units
Feet	0.3048	Meters
Inches	2.54	Centimeters
Inches	0.0254	Meters
Inches	25.4001	Millimeters
Miles	1.6093	Kilometers
Pounds	453.59	Grams
Pounds	0.4536	Kilograms
Metric Units	**Multiplied By**	**US Units**
Centimeters	0.3937	Inches
Grams	0.03527	Ounces
Kilograms	2.2046	Pounds
Kilometers	0.62137	Miles
Meters	3.2808	Feet
Meters	39.37	Inches
Meters	1.0936	Yards
Millimeters	0.03937	Inches

Appendix B

Threat Mine Operations

This appendix is intended to complement the information presented in other manuals on threat obstacle tactics. Commanders should use this information to give added realism to unclassified staff and combined-arms team training. Obstacle employment norms can change with METT-TC factors for a given AO. Therefore, preoperational training on templating, intelligence, recon, and reduction procedures must be based on the best information available before deployment.

MINE OPERATIONS

B-1. Threat formations contain considerable organic minefield-emplacement capability. Threat rapid-mining capability presents a serious challenge to friendly maneuver.

B-2. Armored tracked mine layers known as *GMZs* (*Figure B-1*) dispense mines at a predetermined spacing of 5.5 meters. Three GMZs can lay a 1,200-meter, three-row minefield, containing 624 mines, in 26 minutes. (This does not include the 12- to 15-minute reload and travel times; travel and reload times increase during limited visibility.) Doctrinally, this minefield would be broken into several minefields, each 200 to 300 meters long. Mine-laying helicopters also support threat mine operations. The HIP and HIND-D helicopters carry two or three dispenser pods of AP or AT mines. Artillery-fired SCATMINEs can also support threat mine operations.

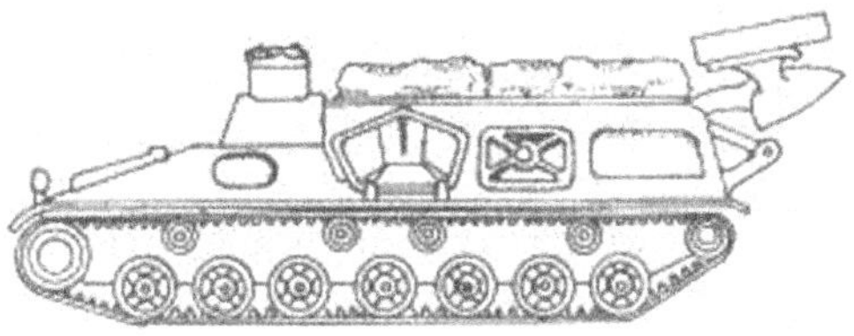

Figure B-1. GMZ Armored Tracked Mine Layer

B-3. Threat armies use obstacles extensively throughout the depth of their defense, and their tactics are chosen well. A threat rapidly emplaced minefield consists of three or four 200- to 300-meter rows, spaced 20 to 40 meters apart, with mines spaced 4 to 6 meters apart.

B-4. *Table B-1* provides detailed information on standard threat AT and AP minefields. Terrain and tactical situations dictate the actual dimensions and distances of minefields.

Table B-1. Normal Parameters for Threat-Style Minefields

AT Minefields	
Front (situation-dependent)	200 to 300 meters
Depth	40 to 120 meters
Number of rows	3 or 4
Distance between rows	20 to 40 meters
Distance between mines	4 to 6 meters for antitrack mines; 9 to 12 meters for antihull mines
Outlay, normal	550 to 750 antitrack mines per kilometer; 300 to 400 antihull mines per kilometer
Outlay, increased effect	1,000+ antitrack mines per kilometer; 500+ antihull mines per kilometer
Probability of destruction	57 percent for antitrack mines (750 per kilometer); 85 percent for antihull mines (400 per kilometer)
AP Minefields	
Front (situation-dependent)	30 to 300 meters
Depth	10 to 150 meters
Distance between rows	5+ meters for blast mines; 25 to 50 meters for fragmentation mines
Distance between mines	1 meter for blast mines; 50 meters (or twice the lethal radius of fragmentation) for fragmentation mines
Outlay, normal	2,000 to 3,000 HE/blast mines per kilometer; 100 to 300 fragmentation mines per kilometer
Outlay, increased effect	2 to 3 times the normal outlay
Probability of destruction	15 to 20 percent for HE/blast mines (2,000 per kilometer); 10 to 15 percent for fragmentation mines (100 per kilometer)

B-5. *Figure B-2* shows a standard rapidly emplaced minefield. The threat army typically uses such a minefield when they are in a hasty defense (offense is temporarily stalled).

B-6. *Figure B-3* shows a standard antitrack minefield.

B-7. *Figure B-4*, page B-4, shows a standard antihull minefield.

B-8. *Figure B-5*, page B-4, shows a standard AP minefield.

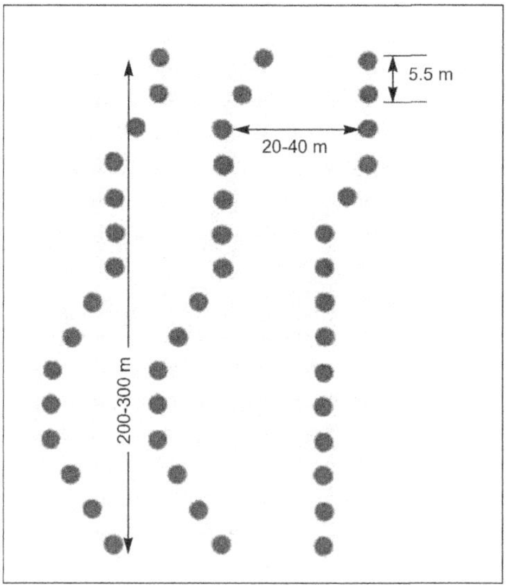

Figure B-2. Threat-Style Rapidly Emplaced Minefield

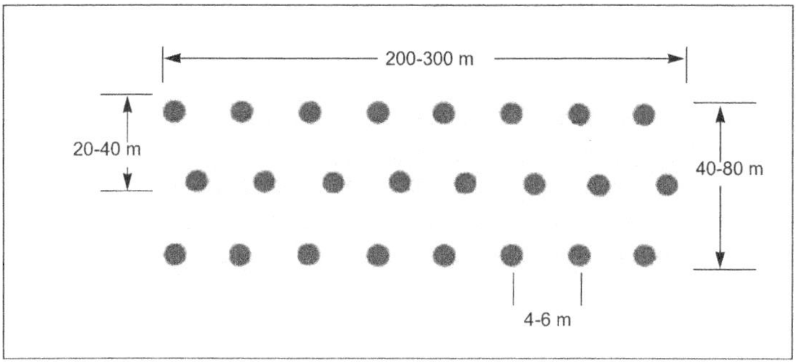

Figure B-3. Threat-Style Antitrack Minefield

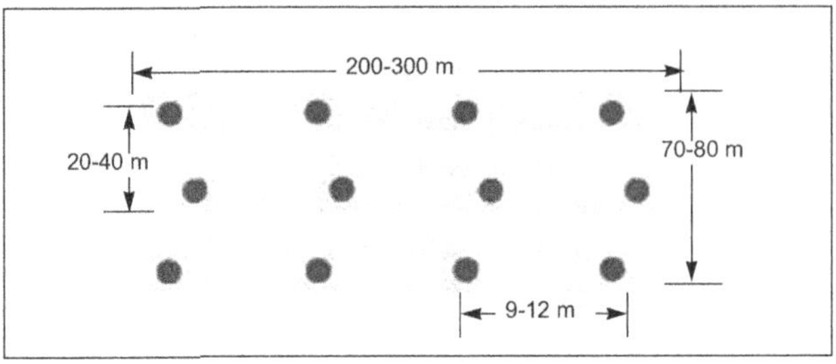

Figure B-4. Threat-Style Antihull Minefield

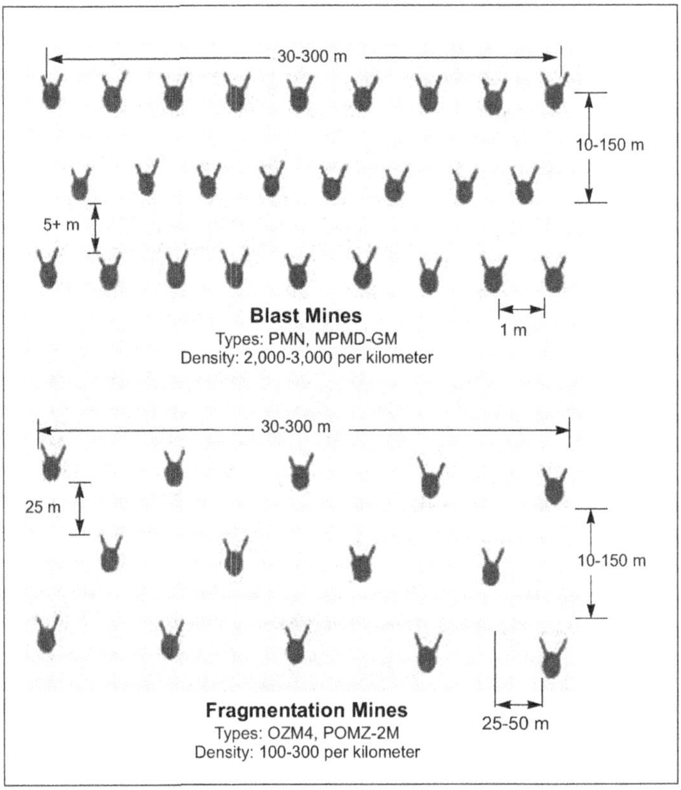

Figure B-5. Threat-Style AP Minefield

B-9. Threat armies also emplace mixed minefields. They are not the same as US mixed minefields. Threat armies normally emplace three rows of AT mines, then several rows of AP mines. AT and AP mines are not normally mixed in the same row.

B-10. Threat engineers use two fundamental drills to emplace mines:

- When emplacing armed mines, the threat uses a crew of five sappers. The first crew member (the senior man and operator) is in the mine layer's seat and monitors the operation of the mine layer and the motion of the mines in the guide chute. He also sets the mine spacing and controls the actions of the GMZ. The second and third members take the mines out of the containers and place them in the intake chute at intervals between the guide tray's drive chain. The GMZ driver steers the vehicle along the indicated route at the established speed.
- When emplacing unarmed mines, the threat uses two or three additional sappers to arm the mines. After emplacing the mines, one sapper trails the mine layer, marks emplaced mines with pennants, and partially camouflages the mines. The remaining sapper(s) then arm the mines.

B-11. Special precautions are taken when emplacing AP minefields. Threat doctrine only allows PMN mines to be surface-laid from mine layers. POMZ-2M mines are emplaced with the truck-and-tray technique. Extra effort is required to assemble, emplace, and deploy the trip wire and to camouflage the POMZ-2M mine.

B-12. Threat forces may also have ground-emplaced SCATMINE capability. One such system is the UMZ SCATMINE system (*Figure B-6*). The UMZ consists of six firing modules mounted on the back of a Zil-131 truck. Each module has 30 firing tubes, for a total of 180 firing tubes per system. Depending on the type of minefield desired, the UMZ can lay 180 to 11,520 mines without reloading. The UMZ can launch an AT or AP minefield 30 to 60 meters from the vehicle while the truck is driving 10 to 40 kilometers per hour (kph). It takes two men 1½ to 2 hours to reload the UMZ. One UMZ can lay a three-row minefield, 150 to 1,500 meters long, depending on the type of mine that is used.

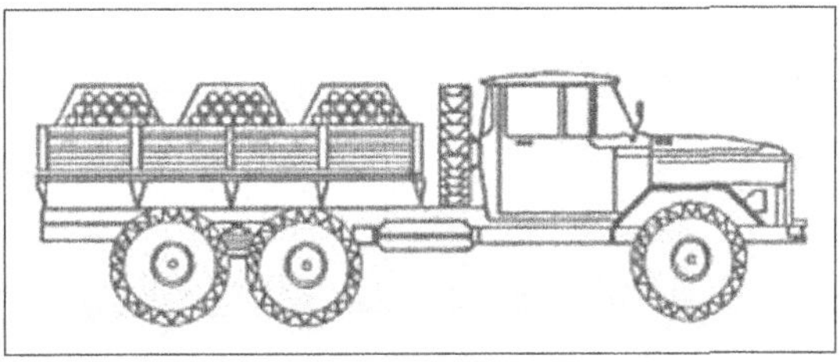

Figure B-6. UMZ SCATMINE System

B-13. For hand-emplaced SCATMINEs, there is a man-portable SCATMINE dispenser. The PKM weighs 2.63 kilograms (without the mine canister) and consists of a single launch tube with a base mount, a blasting machine, and a reel of electric ignition wire. The operator loads a propelling charge and a mine canister into the launch tube and mounts the tube on the edge of a trench or firing parapet. He then aims the tube, connects the ignition wire to the tube, and moves to a safe distance. At an initiating point, the operator connects the ignition wire to the blasting machine and initiates the system. The PKM propels the canister 30 to 100 meters, depending on the type of mine. It lays an AP minefield that is 10 by 20 meters (POM-1S or PFM-1S mine canister) or 10 by 40 meters (POM-2S mine canister). It takes a trained operator 5 minutes to set up the PKM and create a minefield. The PKM can also be used to launch the PTM-1S and PTM-3 AT mine canisters.

B-14. Threat forces use the PKM to lay minefields that protect subunit positions and flanks and the boundaries between subunits. PKM-laid minefields also cover firing lines and gaps in combat formations. The PKM can quickly close breaches in existing minefields and increase the density of mines on armor AAs.

B-15. The type and complexity of an obstacle depend on the installing unit. Maneuver and artillery soldiers usually install simple single-system minefields that are protective in nature. Engineer soldiers install complex obstacles that can include AHDs. Engineer obstacle placement is usually equipment-intensive. Threat engineer effort generally concentrates on tactical obstacles unless maneuver soldiers are unable to employ the necessary protective obstacles. Threat units continue to improve obstacles, supporting their positions by marking the friendly side of the obstacles, burying mines, and adding AHDs.

FOREIGN-MINE DATA

B-16. *Tables B-2 through B-7*, pages B-7 through B-11, provide details on foreign AT and AP mines and mine delivery systems.

Table B-2. Foreign Track-Width AT Mines

Origin	Mechanically Planted	Scatterable	Metallic	Nonmetallic	Simple Pressure	Double Impulse	Blast Hardened	AHD
Russia	TM-46 TM-57 TM-62	PTM-1	TM-46 TM-57 TM-62	TM-62B TM-62D TM-62P	TM-46 TM-57 TM-62		TM-46 TM-57	TMN-46
China	Type 72	Type 84B	Type 72	Type 72	Type 84B	Type 72	Type 72	
Yug	TMM-1		TMM-1	TMA-1 TMA-2 TMA-3 TMA-4 TMA-5	TMA-1 TMA-2 TMA-5		TMA-3 TMA-4	TMM-1 TMA-1 TMA-2 TMA-3 TMA-4 TMA-5
Czech	PT-MI-K PT-MI-BA III		PT-MI-K PT-MI-D	PT-MI-BA II PT-MI-BA III	PT-MI-BA III PT-MI-D		PT-MI-K PT-MI-BA II	PT-MI-K PT-MI-D
Italy	VS-1.6 VS-2.2 TC-.6/6 SB-81	VS-1.6 SB-81		VS-1.6 VS-2.2 TC-.6/6 SB-81			VS-1.6 VS-2.2 TC-3.6 TC-6 SB-81	VS-1.6 VS-2.2
France	M-51 M-52			M-51 M-52	M-51 M-52			M-51
UK	L9 Barmine MK-7		MK-7	L9 Barmine	L9 Barmine MK-7	L9 Barmine MK-7		L9 Barmine
Germany	PM-60 DM-11		DM-21	PM-60 DM-11	PM-60 DM-11 DM-21			PM-60 DM-11 DM-21

Table B-3. Foreign Full-Width AT Mines

Origin	Mechanically Planted	Scatterable	Tilt Rod	Magnetic Influence	Seismic Influence	Coupled	AHD
Russia	TM-62 TM-72 TM-89	PTM-3	TMK-2	TM-62 TM-72 TM-89	TM-62 TM-72 TM-89		TM-62 TM-72 TM-89
China			Type 84C	Type 84A			Type 84A
Yug	TMRP-6	KB-2	TMRP-6	TMRP-6 KB-2			TMRP-6 KB-2
Czech			PT-MI-P				
France	HPD series	Lance	M-51 M-52 Model 1954	HPD series		Model 1953	HPD series M-51 Lance
Hungary	UKB-63		UKB-63				UKB-63
Italy	VS-HCT VS-HCT-2 SB-MV	VS-SATM1 VS-HCT		VS-HCT VS-HCT-2 SB-MV VS-SATM1		Fuse option	VS-HCT VS-HCT-2 SB-MV VS-SATM1
Sweden	FFV-028			FFV-028		Fuse 15	FFV-028
UK	L9 Barmine	HB-876	L9 Barmine MK-7	L9 Barmine HB-876			L9 Barmine MK-7 HB-876
Germany	DM-31	AT-2 MIFF	AT-2	DM-31 MIFF			DM-31 AT-2 MIFF
Romania						MC-71	

Table B-4. Foreign Pressure-Fused AP Mines

Origin	Scatterable	Low Metal	Simple Pressure	Small Diameter	Low-Operating Threshold	Pronged Fuse	Blast Hardened	AHD
Russia	PFM-1		PMN PMD-6		PMN PMD-6 PFM-1	OZM series	PMN-2	
China	GLD-111	Type 72A GLD-111	Type 72A Type 58 GLD-111	GLD-111	Type 72 Type 58 GLD-111			Type 72A Type 72B
Yug		PMB-1 PMB-2 PMB-3	PMA-1 PMD-1		PMA-1 PMA-3 PMD-1	PROM-1 PROM-2 PMR-3	PMA-2 PMA-3	
Czech			PP-MI-D PP-MI-BA		PP-MI-D	PP-MI-SR PP-MI-SR II		
Italy	TS-50 VS-50 SB-33 VS-MK2	TS-50 SB-33 VS-MK2				Valmara 59 Valmara 69	TS-50 VS-50 SB-33 VS-MK2	VS-50 SB-33 VS-MK2
France		M-59		M-59	M-59		M-59	
UK	Ranger		Ranger	Ranger				
Germany		DM-11	PPM-2		DM-11		DM-11	

Table B-5. Foreign Side-Attack AT Mines

Origin	Command	Break Wire	Influence	AHD
Russia	MON-100/200/500, TM-83	LMG	TM-73/83	
Yug		TMRP-6, COBRA		
France	MIACAH F1	MIACAH F1	MIACAH F1	
UK	L14	L14	AJAX	AJAX
Germany			DM-12	
Sweden		FFV-016, AT-4		

Table B-6. Foreign Trip-Wire/Break-Wire-Fused AP Mines

Origin	Scatterable	Trip Wire	Break Wire
Russia	POM-2	POMZ-2, OZM series, MON series	OZM series, MON series
China		Type 58, Type 59, Type 66, Type 69	
Yug		PROM-1/2/KD, PMR-2/3/4, MRUD	PROM-1/2/KD, PMR-2/3/4, MRUD
Czech		PP-MI-SR, PP-MI-SR II, PP-MI-SK	
Italy	VS-SAPFM3	Valmara 59/69, BM-85, P-25, P-40, VS-DAFM1	
France		M1951/1955, MAPED F1	M1951/1955
UK		PAD	
Germany		DM-31, SM-70	

Table B-7. Foreign Mine Delivery Systems

Origin	System	Type[1]	Range	Mine Payload	Density[2]	Remarks
Russia	BM-21	R	20 km	PTM 1/3, POM-2	1.5	1,000 x 500 m per battery
	BM-22	R	35 km	PTM 1/3, POM-2	1.5	
	BKPI	F, H	NA	PTM 1/3, PFM-1, POM-2	1.5	Lays relatively narrow strips
	KSMU	F	NA	PTM 1/3, PFM-1, POM-2	1.5	
	UMZ	V	NA	PTM 1/3, POM-2	1.5	
	GMZ/PMZ-4	MP	NA	Varies	0.4 to 1.0	3 systems can emplace 1,200 m with 624 mines in 26 minutes
	VMR-1/2	H	NA	Varies	0.4 to 1.0	3 Mi-8s can lay 1,200 to 3,000 m with 1,200 mines
China	Type 83	R	15 km	Type 84	0.5	900 x 800 m per battery
Yug	M87	R	50 km	KB-2	0.5	900 x 500 m per battery

Table B-7. Foreign Mine Delivery Systems (continued)

Origin	System	Type[1]	Range	Mine Payload	Density[2]	Remarks
Italy	FIROS 25	R	22 km	VS-HCL, VS-Mk, VS-SAPFM 3	0.5	
	FIROS 30	R	35 km	VS-HCL, VS-Mk, VS-SAPFM 3	0.5	
	DAT	H	NA	Varies	1.2	Lays relatively narrow strips of up to 320 AT mines
	Istrice	V	NA	SATM1, VS-Mk 2 EL, VS-SAPFM 3	Max 1.6	Typically 360 x 140 m
	ST-AT/V	MP	NA	Varies	0.3 to 1.5	240 mines per hour
	Grillo 90	MD	NA	SATM1, VS-Mk 2 EL	0.5	5 AT or 15 AP mines per dispenser
Sweden	FFV-5821	MP	NA	FFV-028	0.4 to 1.4	300 buried mines per hour or 500 surface-laid mines per hour
France	Minotaur	V	NA	Lance	0.7	1,200 x 600 m with 800 mines
	EBG	V	NA	Lance	1.5	60 x 600 m with 40 mines
	155-mm How	A	18 km	Lance		8 mines per round
	Matenin	MP	NA	HPD series	0.4 to 1.4	500 mines per hour
UK	JP-233	F	NA	HB 876		430 mines per Tornado aircraft
	Barmine layer	MP	NA	L9 Barmine	0.4 to 0.9	Lays up to 700 mines per hour
	Ranger	V	NA	L10		1,296 mines per dispenser
Germany	Skorpion	V, H	NA	AT-2	0.4	1,500 x 50 m with 600 mines
	MW-1	F	NA	MIFF, MUSPA	0.5	55 to 500 m wide and 200 to 2,500 m
	LARS	R	14 km	AT-2	0.5	1,836 AT mines per launcher, 1,000 x 500 m per battery
	MARS	R	30 km	AT-2	0.3	1,000 x 500 m with 336 AT mines per launcher
	FFV-5821	MP	NA	DM-31	0.4 to 1.4	300 buried mines per hour or 500 surface-laid mines per hour

[1] A = artillery, H = helicopter, F = fixed-wing aircraft, R = rocket, V= vehicle dispensed, MD = man-portable dispenser, MP = mechanical planter

[2] AT mines per meter of front

Appendix C

Obstacle-Reduction Techniques

Obstacle reduction is the physical creation of a lane through an obstacle. It is a fundamental of breaching operations as discussed in earlier chapters. A number of tasks (detecting, reporting, reducing, proofing, and marking) directly support or are included in obstacle reduction.

DETECTING

C-1. Detection is the actual confirmation of the location of obstacles. It may be accomplished through recon, or it may be unintentional (such as a vehicle running into a mine). Detection is used in conjunction with intelligence-gathering operations, bypass recons, and breaching and clearing operations. There are four types of detection methods—visual, physical (probing), electronic, and mechanical.

VISUAL

C-2. Visual detection is part of all combat operations. Personnel visually inspect the terrain for the following obstacle indicators:

- Trip wires.
- Signs of road repair (such as new fill or paving, road patches, ditching, and culvert work).
- Signs placed on trees, posts, or stakes. Threat forces mark their minefields to protect their own forces.
- Dead animals.
- Damaged vehicles.
- Disturbances in previous tire tracks or tracks that stop unexplainably.
- Wires leading away from the side of the road. They may be firing wires that are partially buried.
- Odd features in the ground or patterns that are not present in nature. Plant growth may wilt or change color, rain may wash away some of the cover, the cover may sink or crack around the edges, or the material covering the mines may look like mounds of dirt.
- Civilians who may know where mines or booby traps are located in the residential area. Civilians staying away from certain places or out of certain buildings are good indications of the presence of mines or booby traps. Question civilians to determine the exact locations.
- Pieces of wood or other debris on a road. They may be indicative of pressure or pressure-release firing devices (FDs). These devices may be on the surface or partially buried.

- Patterns of objects that could be used as a sighting line. An enemy can use mines that are fired by command, so road shoulders and areas close to the objects should be searched.
- Berms. They may indicate the presence of an AT ditch.

PHYSICAL

C-3. Physical detection (probing) is very time-consuming and is used primarily for mine-clearing operations and self-extraction. Detection of mines by visual or electronic methods should be confirmed by probing. Use the following procedures and techniques when probing for mines:

- Roll up your sleeves and remove your jewelry to increase sensitivity. Wear a Kevlar™ helmet, with the chin strap buckled, and a protective fragmentation vest.
- Stay close to the ground and move in a prone position to reduce the effects of an accidental blast. When moving into a prone position—
  - Squat down without touching your knees to the ground.
  - Scan forward up to 2 meters and to the sides up to 3 meters for mine indicators.
  - Probe the area around your feet and as far forward as possible.
  - Kneel on the ground after the area is found to be clear, and continue probing forward until you are in a prone position.
- Use sight and touch to detect trip wires, fuses, and pressure prongs.
- Use a slender, nonmetallic object as a probe.
- Probe every 5 centimeters across a 1-meter front.
- Push the probe into the ground gently at an angle that is less than 45 degrees.

DANGER
Use extreme caution when probing. If the probe is pushed straight down, its tip may detonate a pressure fuse.

- Apply just enough pressure on the probe to sink it slowly into the ground.
- Pick the soil away carefully with the tip of the probe. Remove the loose dirt by hand if the probe encounters resistance and does not go into the ground freely. Take care to prevent detonating the mine.
- Stop probing and use two fingers from each hand to carefully remove the surrounding soil when a solid object is touched and identify the object.
- Remove enough soil to show the mine type (if the object is a mine) and mark its location. Do not attempt to remove or disarm the mine. Use explosives to destroy detected mines in place or use a grappling hook and rope to cause mines to self-detonate. Do not use metal grappling hooks on magnetic-fused mines.

C-4. Probing is extremely stressful and tedious. The senior leader must set a limit to the time a prober will actually probe. To determine a reasonable time, the leader must consider METT-TC factors, weather conditions, the threat level, the unit's stress level, and the prober's fatigue level and state of mind. As a rule, 20 to 30 minutes is the maximum amount of time that an individual can probe effectively.

ELECTRONIC

C-5. Electronic detection is effective for locating mines, but this method is time-consuming and exposes personnel to enemy fire. In addition, suspected mines must be confirmed by probing.

C-6. The AN/PSS-12 mine detector (*Figure C-1*) can only detect metal, but most mines have metal components in their design. The detector can locate and identify plastic or wooden mines by a slight metallic signature. Employment and operation procedures for the AN/PSS-12 are discussed in *FM 20-32*, and technical data is available in *Technical Manual (TM) 5-6665-298-10*. The detector is handheld and identifies suspected mines by an audio signal in the headphones.

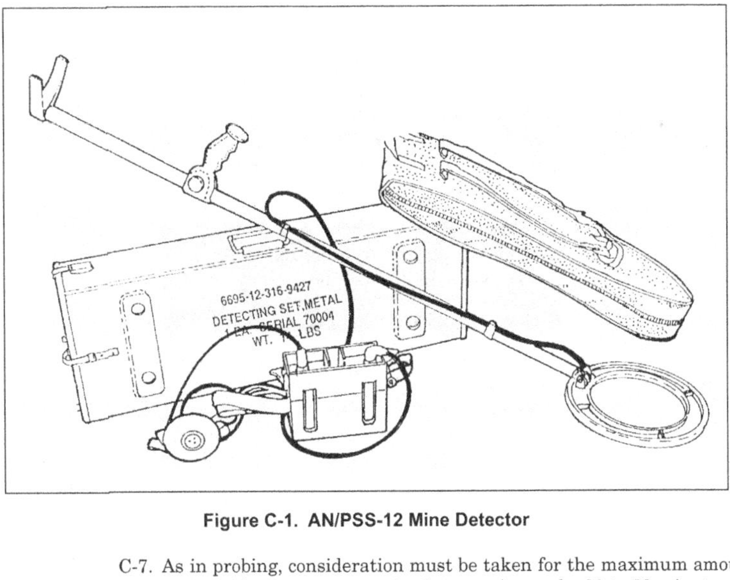

Figure C-1. AN/PSS-12 Mine Detector

C-7. As in probing, consideration must be taken for the maximum amount of time an individual can operate the detector. As a rule, 20 to 30 minutes is the maximum amount of time an individual can use the detector effectively.

MECHANICAL

C-8. The track-width mine roller is a mechanical minefield-detection system. It is most effectively deployed to lead columns on route movement, but it can be used to precede tactical formations. In column movement, unit vehicles travel a narrow path and one or two mine rollers can effectively detect mines in the path. Mine rollers can also be used to detect minefields in front of deployed tactical formations, although more than one roller is required for a good probability of detection.

REPORTING

C-9. Intelligence concerning enemy minefields is reported by the fastest means available. Spot reports (SPOTREPs) are the tactical commander's most common source of minefield intelligence. They originate from patrols that have been sent on specific minefield recon missions or from units that have discovered mine information in the course of their normal operations. The information is transmitted to higher HQ.

REDUCING

C-10. Obstacle-reduction and clearing equipment is broken down into explosive, mechanical, electronic, and manual. Combat engineers and the operators of reduction assets practice and become proficient in these reduction means. They integrate them into breach drills of units they support. The team applies different tactics, techniques, and procedures (TTP) to breach drills and prepares and rehearses them as part of the TF plan. *Tables C-1 through C-4*, pages C-5 through C-11, provide a summary of the effectiveness of reduction assets against the various types of obstacles that may be encountered.

Table C-1. Mechanical Reduction Assets Versus Threat Obstacles

Asset	Track-Width AT Mine	Full-Width AT Mine	Side-Atk AT Mine	Cmd-Det Blast Mine	AT Ditch and Other Gaps	Wire Obstacles	Remarks
IDA	Is more than 90 percent effective to 30.4 cm deep at speeds less than 10 kph (soil dependent)	Is effective against magnetic-influence mines; is more than 99 percent effective against tilt-rod mines	Do not use	Do not use	May be effective at filling in some shallow gaps or trenches	Do not use	The MCB provides almost 100 percent clearance of simple pressure-fused AT mines when used with the MICLIC. The MCB has minimal impact on the M1's maneuverability and speed when not plowing. The main gun should be traversed to the side to prevent damage in case the mine detonates when plowing. The point of breach should be relatively flat and free of large rocks and other obstructions. Plowing operations should begin about 100 m short of the suspected minefield and continue 100 m past its suspected end (typically 2 to 2½ minutes to cover 300 to 400 m overall). Plowing operations should be performed in a straight line to keep the vehicle out of uncleared ground. Fifteen percent of pressure-fused AT-mine encounters across the MCB's width can be expected to detonate under the skid shoe, probably disabling the blade. The MCB is vulnerable to coupled mines and may be stopped by log-hurdle obstacles laid within a minefield as part of a complex obstacle. A second plow should not be used to clear the center of the lane because this will simply push mines into the area cleared by the first plow.

Table C-1. Mechanical Reduction Assets Versus Threat Obstacles (continued)

Asset	Track-Width AT Mine	Full-Width AT Mine	Side-Atk AT Mine	Cmd-Det Blast Mine	AT Ditch and Other Gaps	Wire Obstacles	Remarks
MCR with IDA	Is more than 90 percent effective to 30.4 cm deep at speeds less than 10 kph (soil dependent)	Is effective against magnetic-influence mines; is more than 99 percent effective against tilt-rod mines	Do not use	Do not use	Do not use	Is effective against barbed wire	Multiple rollers are required to provide the overlap necessary for proofing the route's entire width when the MCR is used for proofing during route-clearance operations. Vehicle speed is limited to 5 to 15 kph. The main gun should be traversed to the side to prevent damage in case the mine detonates when rolling a suspected mined area. The IDA may be damaged by detonated mines.
AVLB	NA	NA	NA	NA	Can cross up to MLC 60 (70 if improved) traffic over gaps up to 17.4 m (unprepared abutments) and 18.3 m (prepared abutments)	NA	Is carried on a launcher (modified M48A5 or M60A1 chassis) Weighs 15 tons (bridge only) Can be launched in 2 to 5 minutes
Wolverine	NA	NA	NA	NA	Can cross MLC 70 traffic over gaps up to 24 m	NA	Is carried on a launcher (modified M1-series Abrams tank chassis) Weighs 12.5 tons (bridge only) Can be launched in less than 5 minutes
ACE	NA	NA	NA	NA	Is very effective	NA	

Table C-1. Mechanical Reduction Assets Versus Threat Obstacles (continued)

Asset	Track-Width AT Mine	Full-Width AT Mine	Side-Atk AT Mine	Cmd-Det Blast Mine	AT Ditch and Other Gaps	Wire Obstacles	Remarks
MiniFlail	Is effective against simple pressure-fused AP mines Is less effective against blast-hardened AP mines	Is effective	Is ineffective	Is armored against frag-menting threats Is vulner-able to command-detonated blast mines	NA	Do not use	AT mines can destroy the MiniFlail. Commanders must weigh the use of the MiniFlail against other reduction assets when conducting minefield-clearance operations. Can clear at the rate of about 1,500 square meters per hour

Table C-2. Manual Reduction Assets Versus Threat Obstacles

Asset	Pressure-Fused AP Mine	Trip-Wire/ Break-Wire-Fused AP Mine	Influence-Fused AP Mine	Cmd-Det Blast Mine	AT Ditch and Other Gaps	Wire Obstacles	Remarks
APOBS	Effectively clears 95 percent, including blast-resistant types	Effectively cuts 99 percent of trip wires in its path	Is ineffective	Is ineffective	NA	Clears a footpath	Provides a 0.6- x 45-m footpath from a 25-m standoff Can be deployed by two men in 2 minutes Consists of two manpack units weighing about 55 lbs each
Bangalore torpedo	Effectively clears 85 percent of simple pressure-fused mines; is less effective against blast-hardened mines	Effectively cuts 99 percent of trip wires in its path	Is ineffective	Is ineffective	NA	Clears a footpath	Provides a 0.6- x 15-m footpath Can be deployed by a squad in 5 minutes Consists of ten 1.5-m-long sections weighing 13 lbs each Is ineffective against AT mines
WLGH	Is ineffective	Effectively clears 99 percent of trip wires/break wires in a single pass	Is ineffective; however, it may snag the wires between the control unit and its mines	Is ineffective; however, it may snag the wires between the observer and his mines	NA	Is somewhat effective at clearing trip-wire-fused mines integrated with wire obstacles	Has a range of 80 to 100 m when launched Can be thrown 15 to 20 m when hand thrown Must consider the lethal range/radius (up to 90 m in some cases) of fragmenting AP mines that may be encountered. Select a covered position or move back an adequate distance before pulling the grapnel. Helps increase confidence by grappling an area several times Is hampered by thick vegetation/overhead obstructions that may reduce the WLGH's range/effectiveness

Table C-2. Manual Reduction Assets Versus Threat Obstacles (continued)

Asset	Pressure-Fused AP Mine	Trip-Wire/ Break-Wire-Fused AP Mine	Influence-Fused AP Mine	Cmd-Det Blast Mine	AT Ditch and Other Gaps	Wire Obstacles	Remarks
Probe	Is effective at locating all buried mines; however, extreme caution should be used when probing for small mines (less than 7.6 cm in dia) and sensitive mines (more than 10 lb of actuating force)	Is ineffective; the soldier must rely on clearance by other means and visual detection	Do not use	Ensure that observer is neutralized before approaching	NA	Is ineffective	
Wire cutter	Is ineffective	Do not cut a slack wire until both ends have been checked if break-wire-fused mines may be present. Do not cut taut trip wires until both ends have been checked for tension-release FDs	Do not use	Ensure that observer is neutralized before approaching	NA	Is effective; sound can be minimized by wrapping the wire with a cloth before cutting it	
Gunfire	Is effective against surface-laid munitions	Is effective against surface-laid munitions	Do not use	Is effective against surface-laid munitions	NA	Is ineffective	Is used as a last resort Is a relatively slow method for clearing visually detected ordnance Ensures protection of weapons operator Works better with larger caliber weapons (up to 25 mm); incendiary ammo works best May be the only option open to artillery, MP, and transportation units for dealing with SCATMINEs in the absence of engineers and reduction assets

Table C-3. Dismounted Breaching Assets Versus Threat Obstacles

Asset	Pressure-Fused AP Mine	Trip-Wire/ Break-Wire-Fused AP Mine	Influence-Fused AP Mine	Cmd-Det Blast Mine	AT Ditch and Other Gaps	Wire Obstacles	Remarks
Demolitions	Place at least a 1-lb charge within 15.2 cm of simple pressure-fused mines Ensure that the charge is placed within 2.54 cm of blast-hardened mines	Place at least a 1-lb charge within 15.2 cm of the mine after the mine at the end of a trip wire has been located Can use elevated charges if necessary against the claymore and stake-type mines	Do not use	Ensure that observer is neutralized before approaching Can use elevated charges if necessary against claymore mines	NA	Is effective when used to make an improvised bangalore torpedo	
Blast overshoes	May be effective against very small (up to 1 oz of explosive) mines Is ineffective against mines with more than 1-oz explosive weight	Can reduce casualty severity against fragmenting mines	Can reduce casualty severity against fragmenting mines	Ensure that observer is neutralized before approaching Can reduce casualty severity	NA	Is ineffective	
IBASIC	Will protect personnel near the individual who detonates a pressure-fused AP mine	Can reduce casualty severity against fragmenting mines	Can reduce casualty severity against fragmenting mines	Ensure that observer is neutralized before approaching Can reduce casualty severity	NA	Is ineffective	

Table C-4. Electronic Detection Assets Versus Threat Obstacles

Asset	Pressure-Fused AP Mine	Trip-Wire/ Break-Wire-Fused AP Mine	Influence-Fused AP Mine	Cmd-Det Blast Mine	AT Ditch and Other Gaps	Wire Obstacles	Remarks
AN/PSS-12	Is very effective at finding metallic mines Is less effective against low-metal mines	Is effective at finding metallic trip wires and the fragmenting mines to which they are normally attached Cannot detect monofilament fishing line used for trip wires; it is undetectable by the AN/ PSS-12. Visual detection and other neutralization means should be used first	Do not use	Ensure that observer is neutralized before approaching	NA	Is ineffective	Is less effective against low-metal mines and in areas contaminated with metal fragments (artillery, mortar, small arms) or areas that have heavily mineralized soils

EXPLOSIVE

C-11. The different types of explosive minefield-reduction and clearing equipment are discussed below.

M58A4 Mine-Clearing Line Charge

C-12. The MICLIC (*Figure C-2*) is a rocket-propelled, explosive line charge. It is used to reduce minefields that contain single-impulse, pressure-activated AT mines and mechanically activated AP mines. It clears a 14- by 100-meter path. The MICLIC has a 62-meter standoff distance from the launcher to the detonation point. The MICLIC's effectiveness is limited against prong AP mines, magnetically activated mines (including some SCATMINEs), top-attack mines, side-attack mines, and mines containing multiple-impulse or delay-time fuses. It also has little effect on other obstacles, such as log and concrete barriers, antivehicular ditches, and walls. The shock effect and the psychological impact of the detonation make the MICLIC a useful weapon in a close fight or in urban areas.

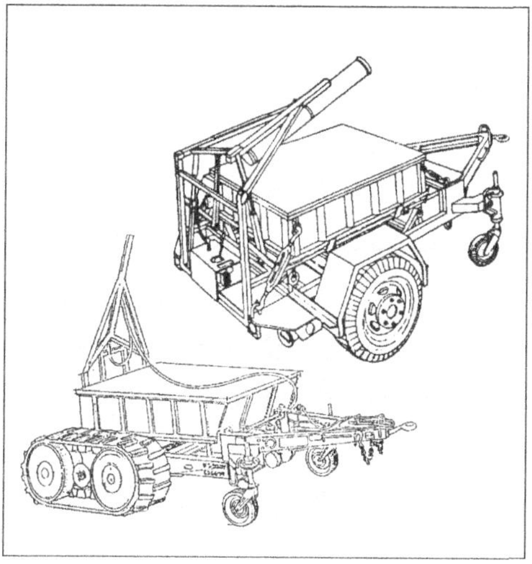

Figure C-2. MICLIC

NOTE: See *Figures C-3* and *C-4* for information on the MICLIC's surface danger zone.

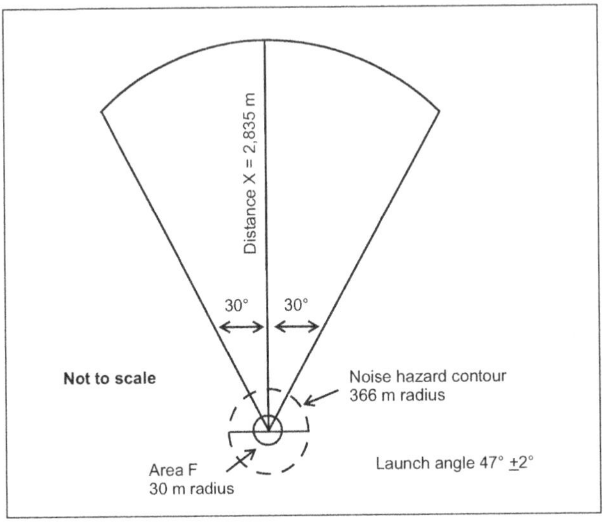

Figure C-3. Surface Danger Zone for Firing the MICLIC

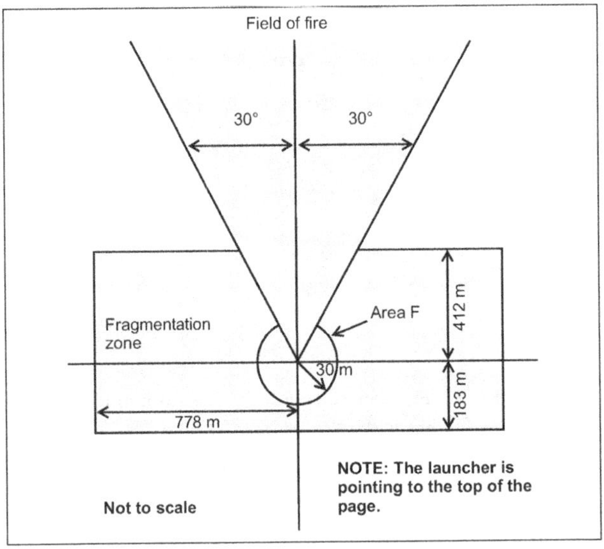

Figure C-4. Area F and Fragmentation Zone (in Detail) for Firing the MICLIC

C-13. The MICLIC is mounted on a rubber-tired trailer, or two MICLICs can be mounted on an AVLB, with the bridge downloaded using a fabricated I-beam frame. Procedures for mounting the MICLIC on the AVLB are outlined in *TM 9-1375-215-14&P*. This is called an armored vehicle-launched MICLIC (AVLM) (*Figure C-5*) and may be the preferred system because no trailer is involved to hinder the mobility of the vehicle.

Figure C-5. AVLM

C-14. Towing vehicles for the trailer-mounted MICLIC are an M113 armored personnel carrier (APC), an M9 ACE, a 5-ton wheeled vehicle, and a 2½-ton wheeled vehicle. The trailer limits the MICLIC's mobility in rough terrain and degrades the maneuverability of the towing vehicle, thereby increasing vulnerability. Since the MICLIC is critical to the breaching operation, it is moved to the point of breach along easily trafficable, covered, and concealed routes. This effectively prevents the towing vehicle from performing any other task (firing or maneuvering) or serving as an engineer squad vehicle unless MICLIC employment is the squad's only mission. This is an important consideration when selecting the towing vehicle because it must be solely dedicated to the mission.

C-15. The MICLIC can be fired from within an armored vehicle without exposing soldiers to fires, although the prefiring preparations must be done at a covered and concealed location near the point of breach. The initiating cable is brought into the vehicle through the hatch, which must be left ajar, or through the portal of the periscope, which has been removed. Therefore, the crew is not afforded NBC protection. When the MICLIC is fired from a wheeled vehicle, the crew must move to a covered position outside the backblast area. The special-purpose cable on the firing control switch is long enough to allow adequate standoff.

C-16. The vehicle operator must be proficient in all aspects of preparing and deploying the MICLIC, including the critical aspect of selecting the optimum point of breach. Although the engineer platoon leader or the breach force commander directs the operator to the point of breach, ensuring that the operator can independently accomplish the task will simplify the operation and greatly enhance its likelihood of success. The towing vehicle and the operator must be selected well in advance and be dedicated solely to the task. The operator must be included in all rehearsals and planning sessions and, if possible, during leader recons. The MICLIC crew and the breaching force must ensure that misfire/malfunction drills are addressed during the unit rehearsal.

C-17. Each MICLIC trailer transports and fires one charge, and then it must be reloaded. The AVLM can fire both MICLICs before reloading. The loaded-charge container weighs 1,283 kilograms, so a lifting device such as a 5-ton wrecker or a heavy expanded mobility tactical truck (HEMTT) is needed. Reloading, which can be done by an experienced crew in about 20 minutes, entails loading a rocket on the rail and lifting a new charge container onto the launcher. The reloading operation must be done in a covered and concealed location.

C-18. The exact limits and depth of an enemy minefield are seldom known before a breach. This is particularly true when the situation is unclear and the minefield is encountered simultaneously with enemy contact. The first and only indication that a unit is in a minefield may be when a vehicle encounters a mine. The minefield's leading edge may still be an uncertainty because the vehicle could have hit a mine in an interior row. The number of MICLICs needed to clear a single lane through a minefield—

- Less than 100 meters deep is one (see *Figure C-6*, page C-16). The minefield's leading edge is identified and, if possible, confirmed by recon. The MICLIC is deployed from a minimum standoff distance of 62 meters from the minefield's leading edge.
- More than 100 meters deep or of uncertain depth is two or more (see *Figure C-7*, page C-16). If the leading edge cannot be identified, the MICLIC is deployed 100 meters from the possible edge or stricken vehicle. When the first MICLIC is detonated, a second MICLIC moves 25 meters into the first MICLIC's path and fires its charge. This extends the lane an additional 87 meters. Additional MICLICs are used for minefields of extreme depth, and each one moves down the lane 25 meters into the path created by the previous charge.

C-19. The neutralization of mines by blast depends on the peak pressure and the impulse. For the MICLIC, the impulse is at a maximum of 3 meters from the line charge (on both sides) and decreases the closer it gets toward the line charge to a minimum of 1 meter from the line charge. This decrease on impulse causes a *skip zone* (see *Figure C-8*, page C-17). This does not mean that neutralization is equal to zero percent; it means that it is not equal to 100 percent. Mines that are buried deeper than 10 centimeters and located 1 to 2 meters from the line charge have a high probability of not being neutralized.

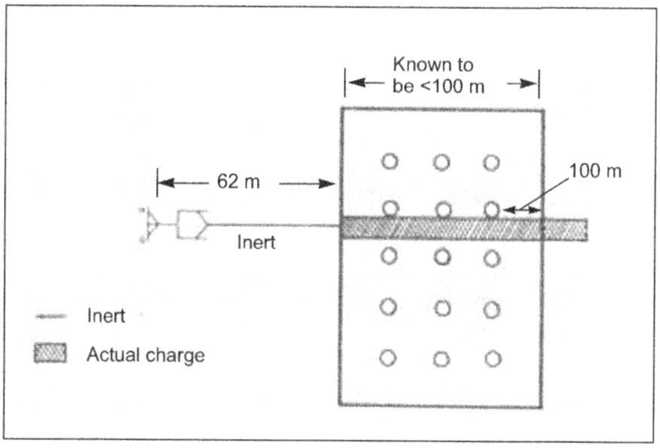

Figure C-6. MICLIC Employment in a Minefield Less Than 100 Meters Deep

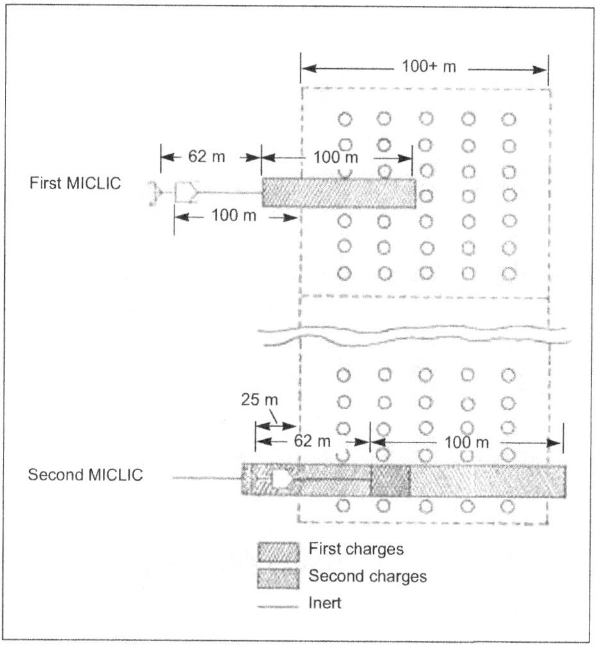

Figure C-7. MICLIC Employment in a Minefield of Uncertain Depth or Greater Than 100 Meters

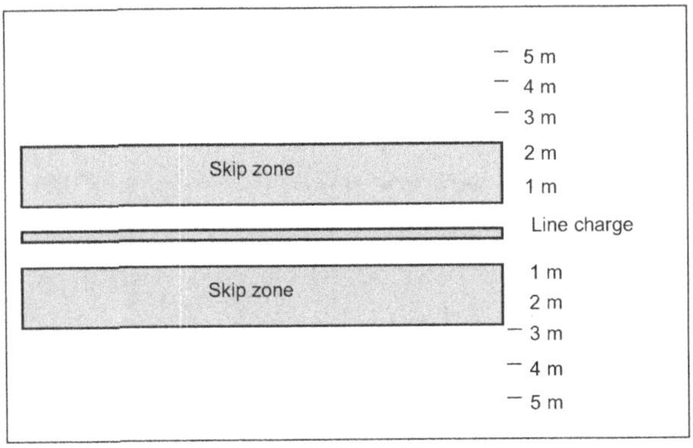

Figure C-8. MICLIC Skip Zone

Antipersonnel Obstacle Breaching System

C-20. The Antipersonnel Obstacle Breaching System (APOBS) (*Figure C-9*) is a man-portable device that is capable of quickly creating a footpath through AP mines and wire entanglements. The APOBS is normally used by combat engineers, infantry soldiers, or dismounted armored cavalry personnel. It provides a lightweight, self-contained, two-man, portable line charge that is rocket-propelled over AP obstacles from a standoff position away from the obstacle's edge.

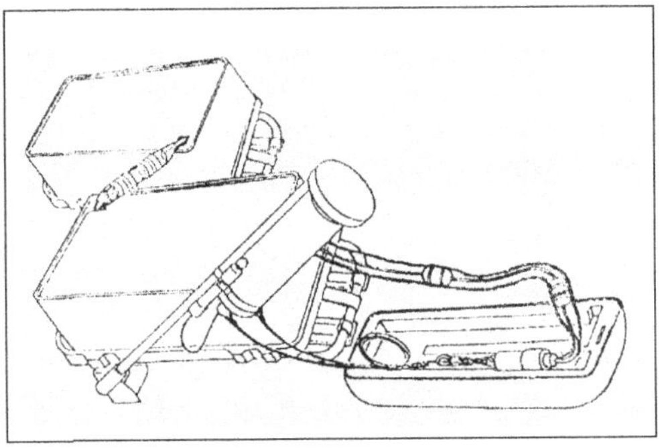

Figure C-9. APOBS

C-21. For dismounted operations, the APOBS is carried in 25-kilogram backpacks by no more than two soldiers for a maximum of 2 kilometers. One backpack assembly consists of a rocket-motor launch mechanism containing a 25-meter line-charge segment and 60 attached grenades. The other backpack assembly contains a 20-meter line-charge segment and 48 attached grenades.

The total weight of the APOBS is about 54 kilograms. It is capable of breaching a footpath about 0.6 by 45 meters and is fired from a 25-meter standoff.

M1A1/M1A2 Bangalore Torpedo

C-22. The bangalore torpedo (*Figure C-10*) is a manually emplaced, explosive-filled pipe that was designed as a wire breaching device, but it is also effective against simple pressure-activated AP mines. It is issued as a demolition kit and consists of ten 1.5-meter tubes. Each tube contains 4 kilograms of HEs and weighs 6 kilograms. The kit clears a 1- by 15-meter lane.

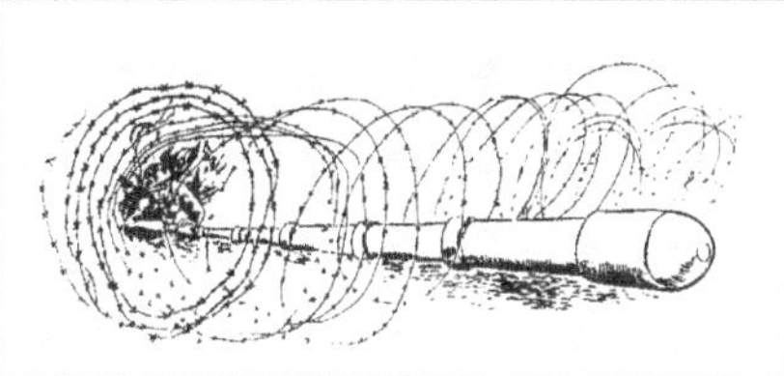

Figure C-10. Bangalore Torpedo

C-23. The bangalore torpedo is used by dismounted infantry and engineer soldiers. An individual soldier or a pair of soldiers connects the number of sections needed and pushes the torpedo through the AP minefield before priming it. A detailed recon is conducted before using the bangalore torpedo to ensure that trip wires have not been used. The bangalore torpedo generates one short impulse and is not effective against pronged, double-impulse, or pressure-resistant AP and AT mines.

WARNING
Do not modify the bangalore torpedo. Cutting the bangalore in half or performing any other modification could cause the device to explode.

MECHANICAL

C-24. MCBs and MCRs are fielded as armor battalion countermine sets that contain 12 MCBs and 4 MCRs. Blades create lanes through minefields, while rollers are used to detect minefields and proof lanes created by other means. Rollers are not a good primary system for lane reduction because multiple mine detonations may destroy the roller system and the vehicle pushing it. (The roller is designed to resist four conventional-mine [22.5 pounds HE] or three SCATMINE strikes, depending on the mine type.)

C-25. The ACE and dozer blades were not designed for reducing minefields and should only be used as a last resort. Using them is extremely time-consuming and hazardous to the crew and equipment. However, ACEs can effectively clear a lane through AP scatterable minefields because they sustain little or no damage and offer protection to the crew. When using a dozer to clear a path through a minefield, the operator is exposed to mine effects. Before clearing begins, sandbag or *up-armor* the operator's cabin and clear the lane of trip wires. When using an engineer blade to clear a path through a scatterable minefield, use the skim technique (*Figure C-11*). Start skimming 100 meters from the suspected leading edge of the minefield.

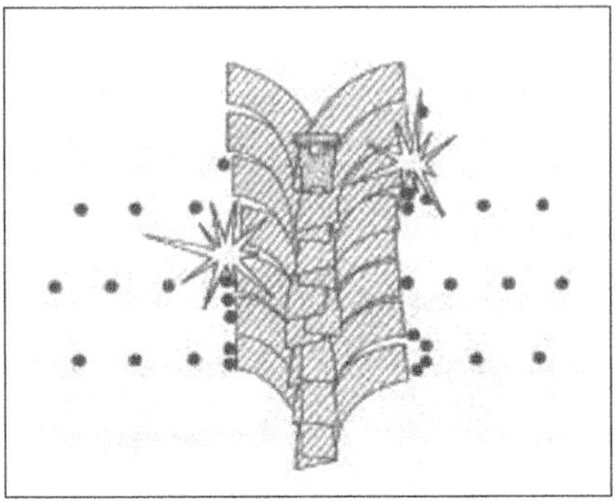

Figure C-11. Skim Technique

C-26. The ACE and dozer blades are very effective against AT ditches and gaps. Normally, during a breaching operation, dozers will not be available due to their inability to keep pace with the attacking forces. Although a single dozer or an ACE is capable of reducing an AT ditch, a technique that has been successfully used and has decreased the time to reduce an AT ditch/gap is to use two ACEs or an ACE/dozer team. When using this technique, position one ACE parallel to the AT ditch and push dirt in front of the second ACE or dozer that is oriented perpendicular to the AT ditch. Use an ACE to push dirt into the AT ditch.

Mine-Clearing Blade

C-27. The MCB (*Figure C-12*, page C-20) is used to remove land mines and create a lane in a minefield. It consists of a blade arrangement with scarifying teeth to extract mines, a moldboard to cast mines aside, and leveling skids to control the blade's depth.

C-28. The MCB lifts and pushes mines, which are surface-laid or buried up to 31 centimeters deep, to the side of track-width lanes. The blade has three depth settings—21, 25, and 31 centimeters. The blade creates a 1.5-meter cleared path in front of each track. *Figure C-13*, page C-20, shows the inside clearance distances between the tracks of common tracked vehicles in relation to the uncleared area left by the MCB.

C-29. Mines armed with AHDs, magnetic fuses, or seismic fuses may be activated when they are lifted by the blade; and they may disable the blade. Mines lifted by the blade are left in the spoil on each side of the furrowed path and remain a hazard until they are removed. Double-impulse mines that are lifted into the spoil on the side can function into the hull of the plowing vehicle.

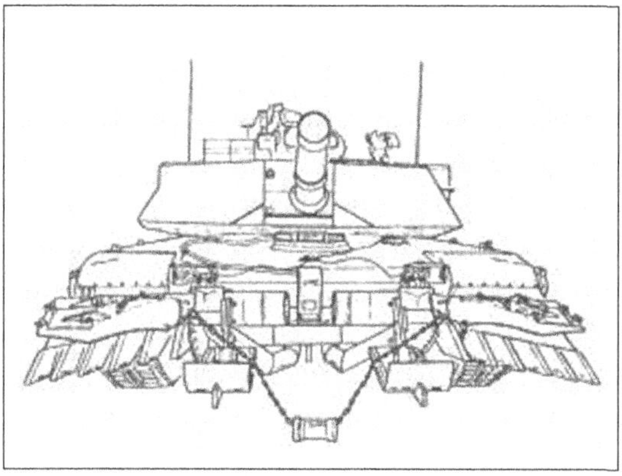

Figure C-12. MCB

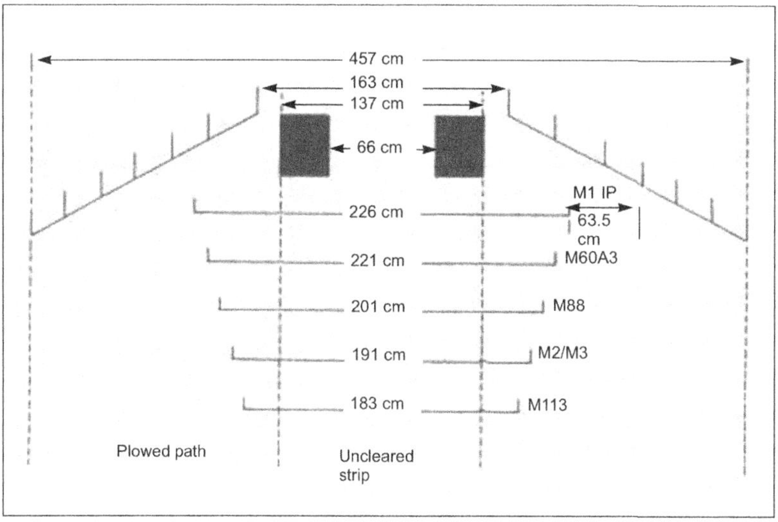

Figure C-13. MCB Width Compared to Tracked-Vehicle Widths

C-30. The skid shoe for each blade exerts adequate pressure to activate most single-impulse mines, which effectively clears a section of the centerline by explosive detonation. This action may disable the blade. Multiple-impulse pressure fuses encountered by the skid shoe are not defeated. A dog-bone assembly between the blades defeats tilt-rod mines. The improved dog-bone assembly (IDA) projects a magnetic signature and defeats tilt-rod and magnetic mines.

C-31. The MCB weighs about 3,150 kilograms and can be mounted on an M1 tank without special preparation or modification. Mounting requires lift capability and takes up to an hour, so it must be mounted well in advance of the mission. It is not easy to mount or transfer the MCB to another tank under battle-space conditions.

C-32. Once mounted, an electric motor raises and lowers the blade. When the blade is raised, it minimally effects the M1's maneuverability and speed. This will not greatly impact the use of the weapon system except when the blade is in operation. The MCB is also equipped with an emergency, quick-disconnect feature.

C-33. The M1 tank crew should maintain a speed of 8 to 10 kph while plowing, depending on soil conditions. The tank cannot maneuver but must continue in a straight path through the minefield to avoid damaging the blade. The main gun must be traversed to the side during plowing because mine detonation under the blade may cause the gun to be thrown violently into the air, damaging the tube. The area selected for the lane must be relatively flat and free of rocks or other obstacles.

C-34. The operator begins plowing about 100 meters from the estimated leading edge of the minefield. He creates a lane extending another 100 meters beyond the estimated far edge of the minefield to ensure that the lane extends through the entire minefield. Multiple vehicles using the lane will deepen the cut made by the MCB, and pressure-fused mines left in the uncleared strip will be dangerous. The uncleared strip should be cleared as soon as possible.

Mine-Clearing Roller

C-35. The MCR (*Figure C-14*, page C-22) consists of a roller assembly, a mounting kit, and a hand winch. The roller assembly weighs about 9,072 kilograms and consists of two push beams mounted to the front of the tank. The rollers are designed to defeat most single-pulse, pressure-activated AT and AP mines. The roller creates a 1.1-meter-wide cleared path in front of each track.

C-36. *Figure C-15*, page C-22, shows the inside clearance distances between the tracks of common tracked vehicles in relation to the uncleared area left by the MCR. An IDA and a chain between the rollers defeats tilt-rod mines. The IDA can be fitted to the roller. The roller is designed to withstand multiple mine explosions before damage; however, this depends on the size of the mines. Large blasts may destroy the roller or the vehicle or injure the crew.

C-37. The roller can be mounted on an M1 or M60 tank that is modified with a permanently attached mine-roller mounting kit. Mounting the roller to a tank is a cumbersome, time-consuming operation because it is very difficult

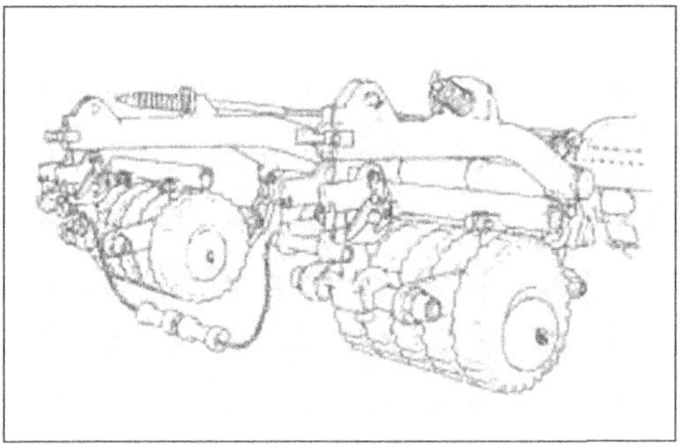

Figure C-14. MCR

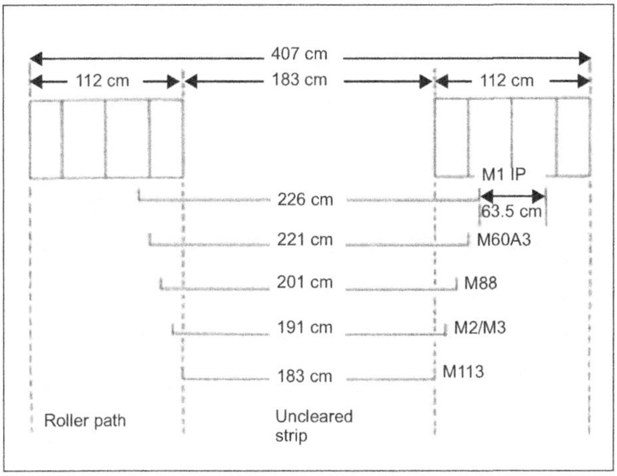

Figure C-15. MCR Width Compared to Tracked-Vehicle Width

under battle-space conditions and requires lift capability. The roller tank is limited to a speed of 5 to 15 kph. When used in a suspected minefield, the MCR must travel in a relatively straight path. Tight turns may cause the roller to deviate from the path of the track and leave the tank vulnerable to mines. Ground fluctuations, bumps, and berms may cause the roller to lift from the ground and miss mines.

C-38. Due to the roller tank's weight and width, units must use caution when crossing an AVLB. In this situation, the curbing from the bridge is removed. To prevent damage to the bridge's hydraulic line, the tank driver uses a strap to lift the IDA and the chain when crossing the bridge. The main gun must be traversed to the side when a mine encounter is possible or imminent. A mine blast can throw the roller or parts of the roller violently into the air and damage the tube. The main gun should only be fired from a temporary halt.

C-39. When the situation and the mission permit, MCRs are used with lead/security elements of the breach force to detect the minefield's leading edge. Rollers are also used to proof lanes in obstacles that are reduced by other means, such as a MICLIC or an MCB. The roller may also be used to lead a supported element traveling in a tactical formation other than a column, but it is less effective than other methods because—

- A vehicle that is not directly behind the roller may encounter mines.
- The roller may travel well into or completely through a widely spaced minefield without encountering a mine, thus giving the formation a false sense of security.
- A mine encountered by the roller may not be on the minefield's leading edge.
- The roller vehicle is extremely vulnerable since it can only use its weapon system from a temporary halt.

C-40. If rollers are to participate in a breaching operation, they must be mounted before rehearsals. Unmounted rollers that are not being used for the mission are carried in the TF formation on M916 tractor trailers. Rollers require lift capability (such as an M88), a secure location, and 30 to 60 minutes to mount on a tank that is fitted with a mounting kit.

Panther

C-41. The M60 Panther (*Figure C-16*) is one of several developmental countermine systems that US forces used during operations Joint Endeavor and Joint TF Eagle. The Panther is a remotely controlled vehicle with mine rollers, and it is used to proof lanes and assembly areas. The system consists of a turretless M60 tank, Israeli mine rollers, an antimagnetic actuating device, and a remote control unit (RCU) that is mounted in a separate vehicle. Additionally, a remote video camera allows the operator to see the road ahead.

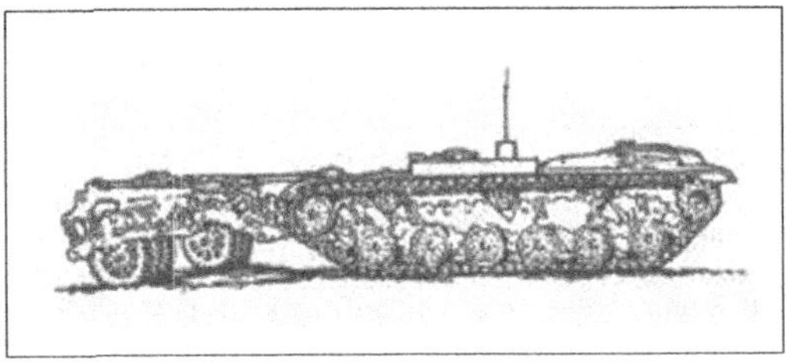

Figure C-16. Panther

C-42. During route-clearance or proofing operations, a Panther is the lead vehicle. It is followed closely by an armored control vehicle, usually an M113. The control vehicle contains the Panther operator, the RCU, and the monitor. The monitor displays the route being proofed or cleared through a camera mounted on the Panther. The Panther is controlled from the commander's cupola or troop hatch of the control vehicle. The control vehicle should be about 200 to 300 meters behind the Panther, and its hatches should be secured open. Crew members in the control vehicle should be wearing Improved Body Armor System, Individual Countermine (IBASIC) protective garments.

C-43. Mine rollers can be raised for limited travel while mounted on the Panther. If the distance is excessive, the rollers must be transported on a cargo carrier. Rollers must be adjusted before every mission to ensure that they have contact with the ground and that their weight is uniformly distributed. Conduct at least three passes to ensure proper coverage and overlap of rollers. Passes should have a minimum of 30 centimeters overlap. Inside roller distances are the same as an MCR.

M1 Grizzly

C-44. The Grizzly (*Figure C-17*) provides the capability for breaching obstacles comprised of mines, wire, posts, rubble, and tank ditches to create a lane for other vehicles to follow. The Grizzly's primary features are a full-width, 4.2-meter MCB and a power arm. The power arm has a reach of 9 meters and a bucket capacity of 1.2 cubic meters. Its primary missions are to reduce berms and fill AT ditches.

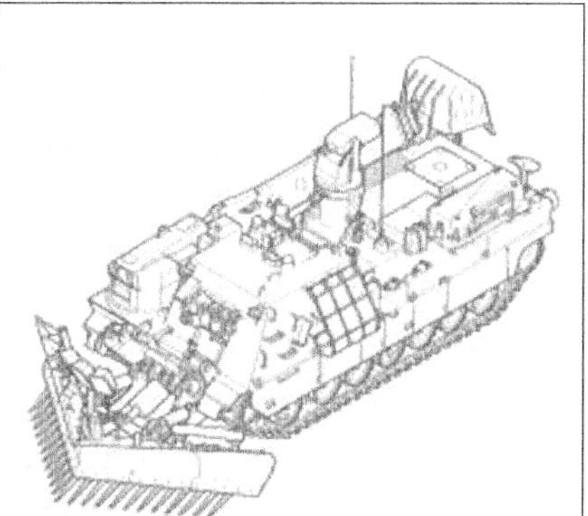

Figure C-17. Grizzly

C-45. The Grizzly lifts and pushes mines, which are surface-laid or buried up to 31 centimeters deep, to the side of full-width lanes. The blade has multiple depth settings, depending on the mission, and it creates a 4.2-meter-wide cleared path. When plowing, the Grizzly is restricted to less than 10 kph, depending on soil conditions. The operator begins plowing about 100 meters from the estimated leading edge of the minefield. He creates a lane extending another 100 meters beyond the estimated far edge of the minefield to ensure that the lane extends through the entire minefield.

C-46. The Grizzly has integrated digital features to enhance battlefield awareness. Some of the digital features are thermal and video cameras, ground-speed sensors, terrain-mapping sensors, and an integrated commander's control station.

Mine-Clearing/Armor-Protection Kit

C-47. The mine-clearing/armor-protection (MCAP) kit consists of two parts—the mine-clearing rake and the armor protection.

- **Mine-clearing rake.** The rake uses a tine that is mounted on a diagonal beam. The rake assembly is designed to sift through the soil, lift out mines, and windrow buried and surface-laid mines to the right of the vehicle. The system clears a 30-centimeter-deep path through a minefield. The rake has a skid shoe that acts as a depth control guide for the operator.
- **Armor protection.** The armor protects the crew against mine blast, small-arms fire, and artillery fire. Protection is also provided for the engine, the fuel tank, and exposed hydraulic cylinders and lines. Ballistic glass blocks are provided at each vision port to permit unrestricted view and operation of the vehicle and the equipment.

The MCAP is mounted on a D7 dozer to perform minefield reduction and lane widening. Proofing the lane must be conducted after the dozer has cleared the lane because some AP mines may still be left in the lane.

Armored Vehicle-Launched Bridge

C-48. The AVLB (*Figure C-18*) is carried on a launcher (a modified M48A5 or M60A1 chassis) and weighs 15 tons. A buttoned-up, two-man crew can launch the AVLB in 2 to 5 minutes. The AVLB can be retrieved from either end; however, one man is exposed to fire while guiding and connecting it. The AVLB is capable of carrying military load class (MLC) 60 (70 with improved bridge) tracked loads across a 17.4-meter gap with unprepared abutments or an 18.3-meter gap with prepared abutments.

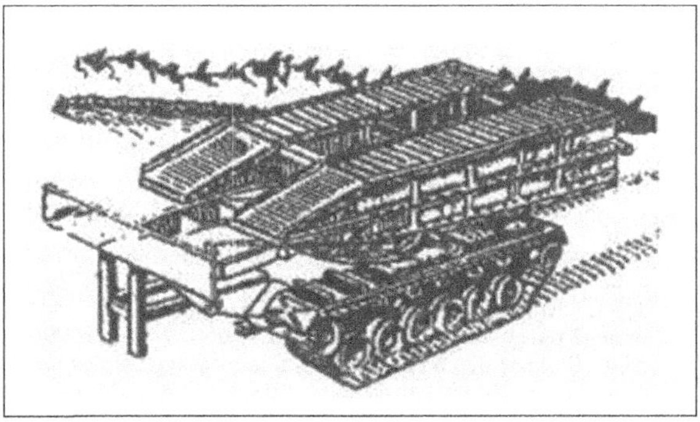

Figure C-18. AVLB

C-49. The AVLB has the following limitations:

- The scissor launch requires 10 meters of overhead clearance.
- The maximum launch slope is—
  - 28 percent uphill.
  - 19 percent downhill.
  - 11 percent on a side slope.

Wolverine

C-50. The AVLB will eventually be replaced by the Wolverine (*Figure C-19*). The bridge is carried on a launcher (a modified M1 tank chassis) and weighs 12.5 tons. It can be launched in less than 5 minutes by a buttoned-up, two-man crew and can be retrieved from either end of the bridge in less than 10 minutes. The Wolverine can cross an MLC 70 tracked vehicle over 24-meter gaps. The launcher has a maximum speed of 83 kph.

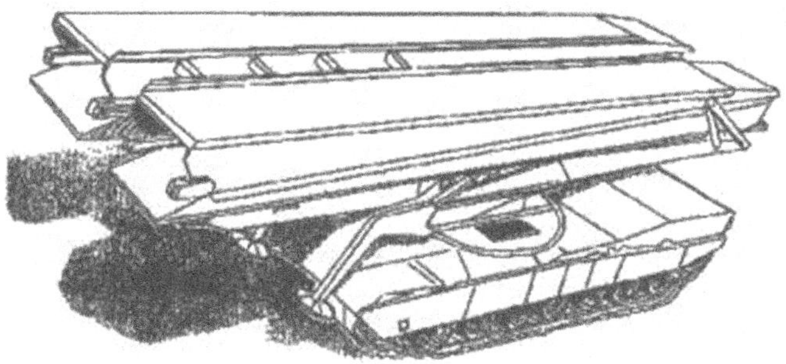

Figure C-19. Wolverine

M9 Armored Combat Earthmover

C-51. The M9 ACE (*Figure C-20*) is capable of performing mobility tasks in support of light and heavy forces. These tasks include gaps and AT ditches. Although an ACE can be used to reduce a scatterable minefield, it should only be used as a last resort.

ELECTRONIC

C-52. The Field-Expedient Countermine System (FECS) is a series of copper coils that fit over the front of tracked and wheeled vehicles. Power is supplied by the vehicle's battery. The coils emit a large magnetic signature that detonates magnetically fused mines located 2 to 5 meters in front of the vehicle. The FECS is designed to defeat magnetically influenced mines only and must be used with other countermine systems.

Figure C-20. M9 ACE

MANUAL

C-53. While mechanical and explosive reduction procedures are normally preferred, a unit may have to use manual procedures for the following reasons:

- Explosive and mechanical mobility assets are unavailable (not on hand or nonmission capable).
- Explosive and mechanical mobility assets are ineffective against the type of obstacle or because of terrain limitations (see *Table C-1*, page C-5).
- Stealth is required.

C-54. Manual procedures involve dismounted soldiers (usually engineers) using simple explosives or equipment to create a lane through an obstacle or to clear an obstacle. These procedures expose the soldier and may be manpower- and time-intensive. However, this method works in most situations and under most conditions. Certain types of terrain, weather, and sophisticated fuses can severely degrade the effectiveness of rollers, plows, and line charges.

Techniques

C-55. As discussed, manual procedures are effective against all obstacles under all conditions. Most types of obstacles do not cause casualties directly, but minefields have the potential to cause direct casualties. The enemy possesses a significant mechanical, mine-burying capability. It has the capacity and the propensity for the labor-intensive effort required to bury mines by hand; however, the enemy often lays mines on the surface. Buried mines are usually found in a highly prepared defense. When training for the reduction of surface-laid and buried minefields, always assume the presence of AHDs and trip wires until proven otherwise. Manual reduction of surface-laid and buried minefields is discussed below:

- **Surface-laid minefield.** Engineers first use grapnel hooks from covered positions to check for trip wires in the lane. The limited range of the tossed hook requires the procedure to be repeated through the estimated depth of the obstacle. A demolition team then moves through the lane. The team places a line main down the center of the lane, ties the line from the explosive into the line main, and places blocks of explosive next to surface-laid mines. After the mines are detonated, the team makes a visual check to ensure that all mines were cleared before directing a proofing roller and other traffic through the lane. Members of the demolition team are assigned special tasks, such as grappler, detonating-cord man, and demolitions man. All members should be cross-trained on all procedures. Demolitions are prepared for use before arriving at the point of breach. An engineer platoon uses squads in series through the minefield to clear a lane for a company team. The platoon must rehearse reduction procedures until execution is flawless, quick, and technically safe. During reduction, the engineer platoon will be exposed in the lane for 5 minutes or more depending on the mission, the minefield depth, and the platoon's level of training.

- **Buried minefield.** Manually reducing a buried minefield is extremely difficult to perform as part of a breaching operation. If mine burrows are not easily seen, mine detectors and probes must be used to locate mines. Mines are then destroyed by hand-emplaced charges. As an alternative, mines can be removed by using a grappling hook and, if necessary, a tripod (*Figure C-21*). Using a tripod provides vertical lift on a mine, making it easier to pull the mine out of the hole.

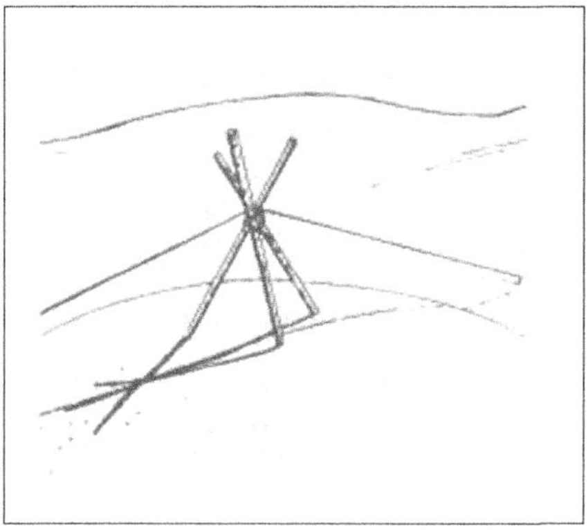

Figure C-21. Tripod

C-56. The platoon leader organizes soldiers into teams with distinct, rehearsed missions including grappling, detecting, marking, probing, and emplacing demolitions and detonating cord. The platoon is exposed in the obstacle for long periods of time.

Grappling Hook

C-57. The grappling hook (grapnel) is a multipurpose tool that is used for manual obstacle reduction. Soldiers use it to detonate mines from a standoff position by activating trip wires and AHDs. After the grapnel is used to clear trip wires in a lane, dismounted engineers can move through the minefield, visually locate surface-laid mines, and prepare mines for demolition. In buried minefields, soldiers grapple, and then enter the minefield with mine detectors and probes. There are two types of grapnel hooks—hand-thrown and weapon-launched.

- **Hand-thrown.** A 60+-meter light rope is attached to the grapnel for hand throwing. The throwing range is usually no more than 25 meters. The excess rope is used for the standoff distance when the thrower begins grappling. The thrower tosses the grapnel and seeks cover before the grapnel and rope touch the ground in case their impact detonates a mine. He then moves backward, reaches the end of the excess rope, takes cover, and begins grappling. Once the grapnel is recovered, the thrower moves forward to the original position, tosses the grapnel, and repeats the procedure at least twice. He then moves to the end of the grappled area and repeats this sequence through the depth of the minefield.
- **Weapon-launched.** A 150-meter lightweight rope is attached to a lightweight grapnel that is designed to be fired from an M16A1 or M16A2 rifle using an M855 cartridge. The grapnel is pushed onto the rifle muzzle with the opening of the retrieval-rope bag oriented toward the minefield. The firer is located 25 meters from the minefield's leading edge and aims the rifle muzzle at a 30- to 40-degree angle for maximum range. Once fired, the grapnel travels 75 to 100 meters from the firer's position. After the weapon-launched grapnel hook (WLGH) has been fired, the firer secures the rope, moves 60 meters from the minefield, moves into a prone position, and begins to grapple. The WLGH can be used only once to clear a minefield, but it can be reused up to 20 times for training because blanks are used to fire it.

C-58. Multiple grapplers can clear a lane of trip wires quickly and thoroughly, but they must time their efforts and follow procedures simultaneously. A hit on a trip wire or a pressure fuse can destroy the hook and the cord, so engineers should carry extras.

Appendix D

Breaching Scenario and Special Planning Considerations

This appendix provides a scenario showing how the breaching tenets may be applied when conducting a breaching operation in open terrain, such as a desert. This example is not an attempt to give a single, doctrinally correct solution but simply to provide a method that a unit may use to task-organize and approach this very difficult mission. Also discussed in this appendix are considerations that may be needed when conducting a breaching operation in restricted terrain (a defile, urban terrain, or the jungle). The following paragraphs are an example of a worst-case scenario in which a TF is given the mission to conduct a breaching operation against a well-equipped, well-prepared enemy defense.

SAMPLE BREACHING OPERATION

D-1. TF 1-12 is a mechanized infantry battalion that is task-organized with two tank companies from TF 3-71 (armored battalion). It just received a mission to secure OBJ SPARROW as part of a brigade attack to destroy a motorized infantry battalion (MIBN) (see *Figure D-1,* page D-2).

D-2. TF 1-12 received intelligence products, including the brigade SITEMP and a series of WOs, before issuing the actual OPORD. The TF staff, in particular the S2 and the engineer, revised and refined the brigade SITEMP based on their analyses of the terrain and the enemy (see *Figure D-2,* page D-3).

D-3. Upon receipt of the brigade OPORD, TF 1-12 learns that—

- The brigade has task-organized it with an engineer recon team to augment its recon capability and its habitually associated engineer company. An additional engineer line platoon has been attached to the engineer company, giving the TF a total of three engineer line platoons.
- The brigade has given TF 1-11 the mission to fix enemy forces north of OBJ SPARROW to prevent them from repositioning.
- The brigade plans to disrupt the enemy's counterattack force by using area-denial artillery munition (ADAM)/remote antiarmor mine (RAAM) SCATMINEs along the high-speed route to OBJ SPARROW. The counterattack force has been located by intelligence sources.
- The brigade plans to have the engineer battalion (-) follow TF 1-12's assault force to assume control of the lanes. The engineer battalion will create additional lanes, if necessary, and upgrade/maintain the lanes and marking pattern to facilitate the passage of TF 3-71.

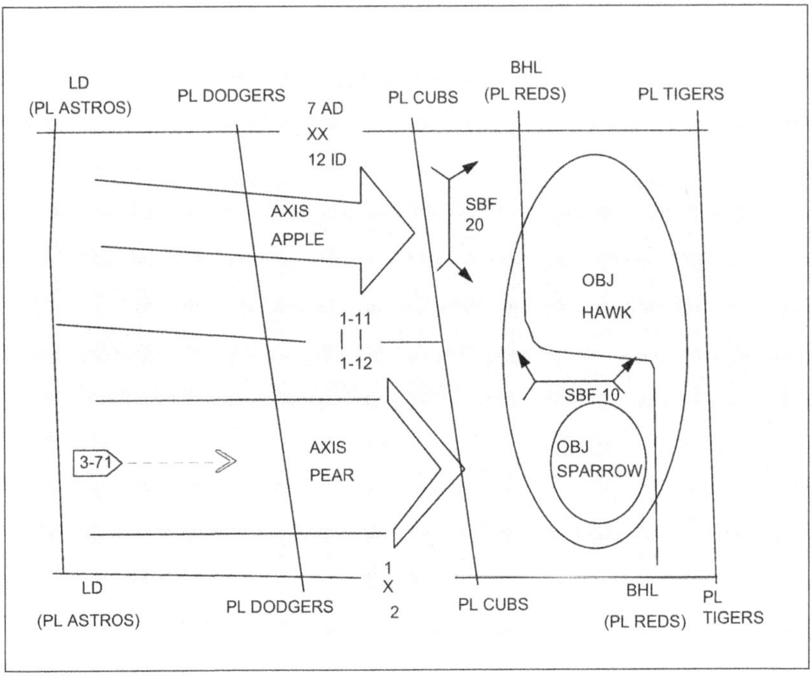

Figure D-1. Brigade Maneuver Graphics

- The brigade has tasked it to seize OBJ SPARROW and occupy SBF 10 with at least two platoons to fix enemy forces north of OBJ SPARROW.
- The brigade has tasked it to create two lanes in the obstacle system to support the passage of TF 3-71 through the created lanes.

D-4. Based on the commander's guidance, the TF staff developed three COAs. The third COA was chosen based on war gaming. It required a company team (assault force), consisting of one tank platoon (the platoon has two tank plows mounted) and two mechanized infantry platoons, to seize OBJ SPARROW. The staff decided that two lanes were necessary in the obstacle system, vicinity grid AA123456, based on the specified task given by the brigade. Because of the lane requirement and the obstacle type expected to be encountered, the staff decided that the breach force must consist of at least two tanks with MCBs, two tanks with MCRs, and two MICLICs. The breach force must also have two engineer platoons to create, proof, and mark the lanes and to provide redundancy of assets. The staff considered the threat to the breach force to be—

- The infantry platoon and AT weapons on the southern flank.

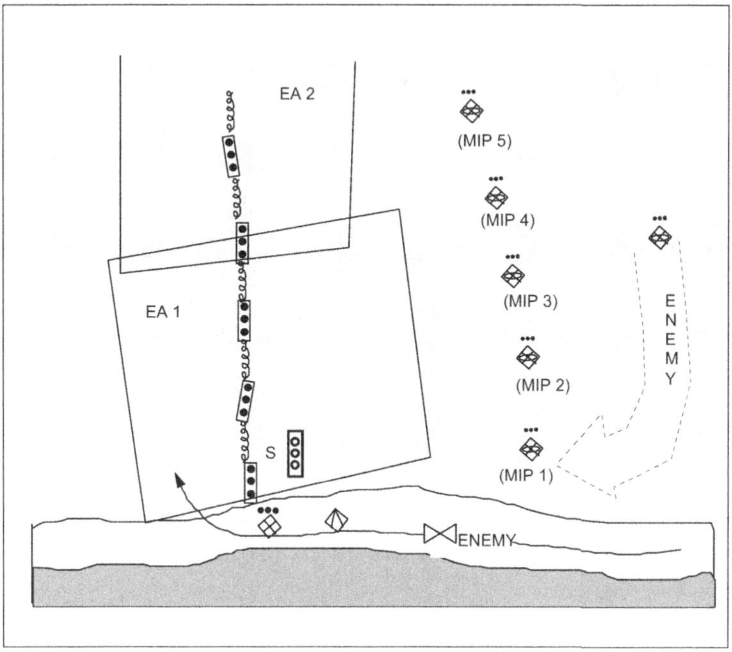

Figure D-2. TF SITEMP

- The southern motorized infantry platoon (MIP), which the support force may not be able to suppress because its forces are masking the breach force.
- The MIP-size counterattack force.

D-5. In this COA, the staff task-organized one tank platoon and one mechanized infantry platoon to the breach force (in addition to the MCBs and MCRs) to provide local security and redundant suppression. The staff determined from its terrain and enemy analyses that the support force's mission is to suppress the four southern MIPs to facilitate reduction operations. To accomplish this, the staff task-organizes two company teams to the support force. Each company team in the support force has one tank plow to reduce any obstacles it encounters en route to the SBFs (see *Figure D-3,* page D-4).

D-6. The staff further refines the COA by developing an initial plan to achieve the following breaching fundamentals, to include assigning tasks to the various subordinate forces (see *Figure D-4,* page D-5):

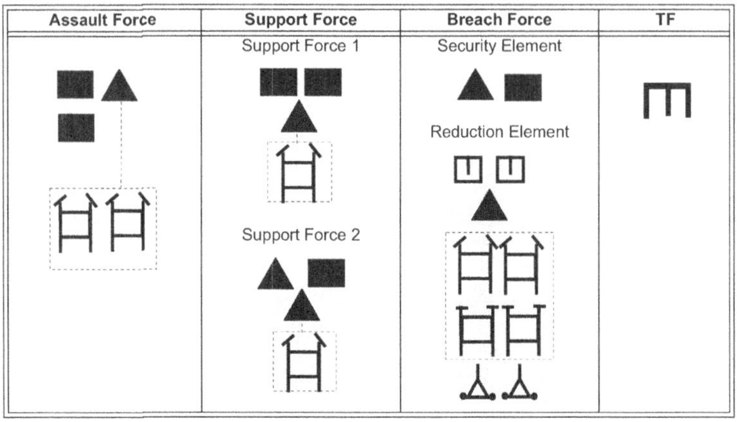

Figure D-3. Initial Breaching Organization

- **Suppress.**
  - Support force 1 will occupy SBF 1 and initially orient its fires on MIPs 1 and 2. When the breach force is committed, it will shift its fires to orient on MIPs 2 and 3. Fires will be lifted on the order of the assault-force commander.
  - Support force 2 will occupy SBF 2 and orient its fires on MIPs 3 and 4. Fires will be lifted on the order of the assault-force commander.
  - The breach-force security element will occupy SBF 3 and orient fires on MIP 1. The security element is responsible for local security at the point of breach.
  - The fire-support element (FSE) will ensure that CFZ 1 is activated as the breach force moves to the point of breach.
  - The support force will call for and adjust artillery targets AG7001 through AG7003 when the lead elements of the support force cross PL CUBS.
- **Obscure.**
  - The support force will call for and adjust smoke target AG1210 when the lead elements of the support force cross PL CUBS. The support force will adjust this target only if the smoke is ineffective.
  - The smoke platoon (task-organized to the TF for this mission) will employ its generators to the north of support force 2. Smoke operations begin when support force 2 crosses PL CUBS.
- **Secure.** The breach force (security element) will occupy SBF 4 and orient fires on the infantry platoon and AT weapons on the south wall.

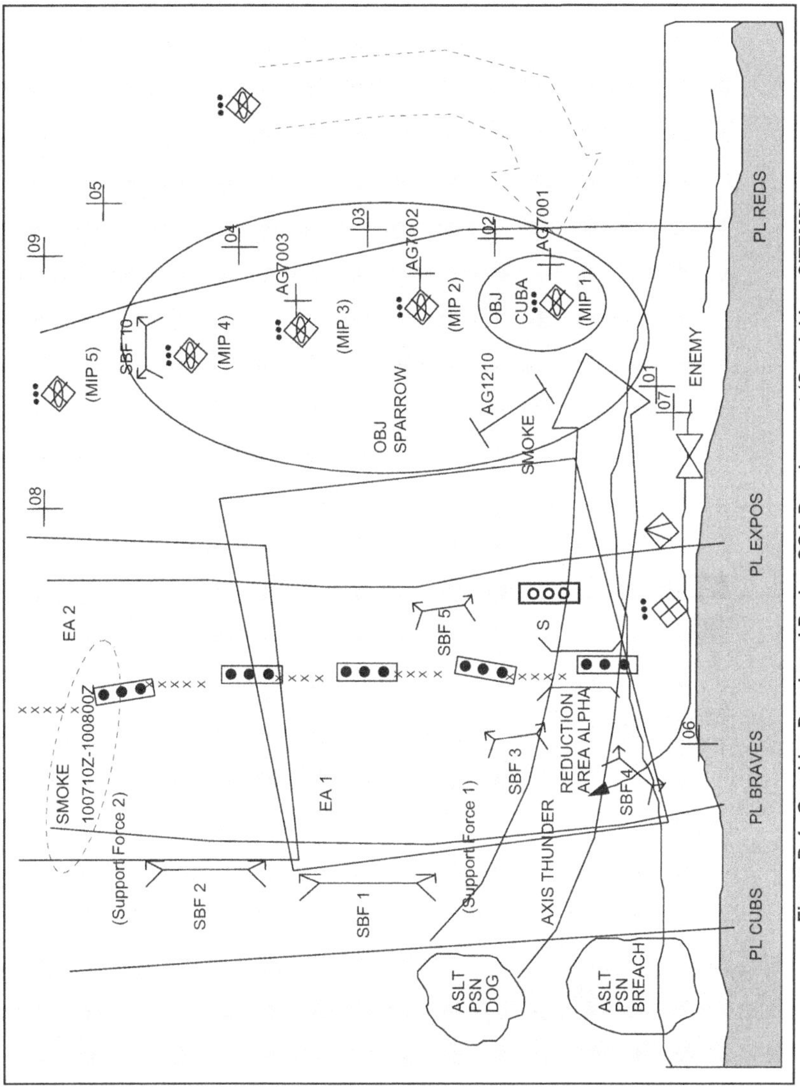

Figure D-4. Graphics Developed During COA Development (Overlaid on SITEMP)

- **Reduce**.
  - The breach-force reduction element will move from assault position (ASLT PSN) BREACH when the commitment criteria in reduction area ALPHA has been met. In this scenario, it will create, proof, and mark (initial lane-marking pattern) two lanes in the obstacle system, at least 800 meters apart, within reduction area ALPHA. This considers the enemy's use of artillery-delivered SCATMINEs to close the lanes.
  - The breach-force security element will occupy SBF 5 with one platoon orienting fires on MIPs 2 and 3 to provide enemy suppression when one lane is created in the tactical obstacles.
- **Assault**.
  - The assault force will attack along axis THUNDER to initially seize OBJ CUBA (point of penetration) and then the rest of OBJ SPARROW.
  - The assault force will occupy SBF 10 with at least two platoons oriented on MIP 5.

D-7. During COA analysis, modifications are made to the original COA. This includes task-organizing an engineer platoon to the assault force to help reduce protective obstacles and mark lanes for the assault and follow-on forces (see *Figure D-5*).

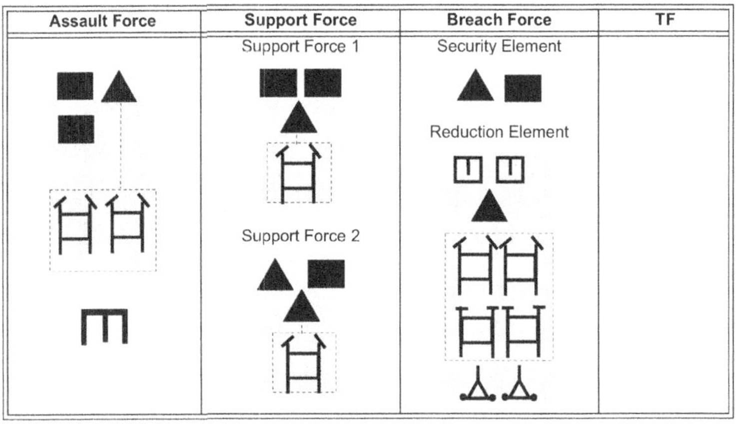

Figure D-5. Final Task Organization

D-8. Additionally, the staff develops a decision point for committing the breach force and a plan to collect the necessary information to support the decision point (see *Table D-1*). *Figure D-6* shows the location of each NAI.

Table D-1. Plan to Support the Decision Point

Decision Criteria	NAI	Responsibility
MIP in OBJ CUBA must be destroyed.	NAI 300	TF scouts
Enemy south of TRP 5 must be suppressed.	NAI 301	Support force commander
Point of breach must be effectively obscured.	NAI 302	Support force commander
Support force must have a minimum of 16 M1s/M2s available.		Support force commander
Breach force must have a minimum of one engineer platoon, 1 MICLIC/ MCB, and 6 M1s/M2s.		Breach force commander
CFZ 1 must be activated.		TOC (FSE)

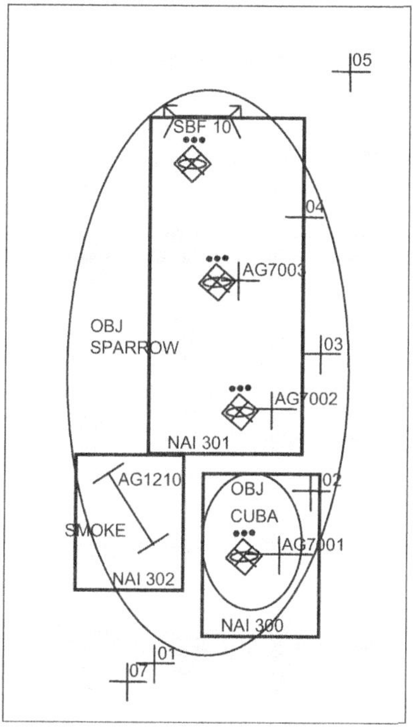

Figure D-6. NAIs to Support the Decision Point

D-9. The commander specifies the following C^2 structure of the mission based on the scheme of maneuver in the COA he approved:

- The assault force will be comprised of Team A/3-71 (mechanized heavy).
- The breach force will be comprised of Team B/3-71 (tank heavy) and A/ 55th Engineer Battalion. The Team B commander will be the breach-force and security-element commander. The engineer commander will be the reduction-element commander.
- The support force will be comprised of Team C/1-12 (mechanized heavy) and Team D/1-12 (tank heavy). The support-force commander will be the TF S3.

D-10. Preparation for the breaching operation involves numerous rehearsals, beginning at the lowest levels and culminating into a full-dress, combined-arms rehearsal. S2 reports that the TF received the confirmed location of the tactical obstacles in the vicinity of the point of breach. The report indicated that the minefields are comprised of three rows of buried SB-MV mines (Italian AT mines), with a uniform 4.5 meters spacing between mines and 70 meters between rows. One row of concertina wire is positioned 40 meters to the rear of each tactical minefield. Additionally, the enemy has emplaced protective minefields, which are comprised of AP mines only, 400 meters in front of each platoon position. The enemy has tanks and infantry fighting vehicles (IFVs) overwatching tactical and protective obstacles. Reports indicate that the enemy has not prepared dismounted positions.

NOTE: The assault-force commander decides to conduct a mounted assault based on information from the intelligence report and the terrain analysis he conducted. The terrain analysis indicates that tank plows will effectively reduce the enemy's protective obstacles.

D-11. This information is quickly relayed to subordinate units and all rehearsal sites reflect these obstacles. The commander pays particular attention to the status and location of reduction assets throughout preparations for the attack. This results in an MCB being removed from an inoperable M1 at the unit maintenance collection point (UMCP) to an operable M1 in the assault force.

D-12. The TF crosses the LD at 95 percent strength (all MCBs, MICLICs, and MCRs are operational). Before crossing PL DODGERS, the TF scouts report that three of the four vehicles in NAI 300 have been destroyed by Copperhead munitions. The commander reminds the S3 that destroying the remaining vehicles is his first priority.

D-13. As the support force crosses PL CUBS, the TF S3 reports that he has lost two M2s to indirect fires and one M1 to maintenance and is beginning to occupy SBFs 1 and 2. Artillery units begin suppressive fires and begin to build obscuration smoke. As the support force begins to suppress the enemy with direct and indirect fires, the breach and assault forces occupy ASLT PSNs BREACH and DOG, respectively, and conduct final preparations as necessary. The tactical operations center (TOC) informs the commander that CFZ 1 will be activated when the breach force is committed. The breach-force commander reports that all reduction assets are operable, and the TF scouts report that the final vehicle in OBJ CUBA has been destroyed. The TF S3 reports that obscuration and suppression have been achieved and that combat power is 6 M1s and 13 M2s.

D-14. Having met the commitment criteria, the commander orders the breach force to AA123456 to begin obstacle reduction. The breach-force commander initially moves his security element to SBFs 3 and 4. The reduction-element commander also moves forward with his platoon leaders to recon the obstacle and confirm/refine the plan to reduce the obstacle. Once the security element is set in SBFs 3 and 4, the reduction element moves forward to reduce the obstacle. When the first lane is created and proofed, the breach-force commander moves a portion of the security element from SBF 3 to SBF 5. The assault-force commander begins movement from the assault position to the far-recognition marker that has been reported by the reduction-element commander. Upon reaching the far-recognition marker, the assault-force commander can observe the final-approach marker and the marked lane.

D-15. The assault force moves through lanes in the tactical obstacles and sets its platoons in the SBFs. It immediately begins to create two lanes in the protective obstacles with the MCBs. When the lanes are complete, the assault-force commander maneuvers the two tank platoons to SBF 10, while the mechanized company team continues to suppress the remaining enemy vehicles. The engineer platoon remains at protective obstacles to mark lanes in preparation for passing follow-on forces.

D-16. The engineer battalion S3 is immediately behind the assault force. He takes control of the reduction element that has created, proofed, and marked two lanes in the tactical obstacles by this time. The S3 moves a portion of the reduction assets away from the lanes to counter any SCATMINEs that the enemy may use to close the lanes. The S3 prepares to assist with TF 3-71's passage through the lanes. When the assault force reports that it has created a safe route to PL REDS, the brigade commander orders TF 3-71 to pass through TF 1-12 and continue the attack.

CONSIDERATIONS FOR RESTRICTED TERRAIN

D-17. The breaching tenets discussed in *Chapter 1* apply to breaching operations in restricted terrain; however, the application of these tenets is different from a breaching operation conducted in open terrain. The following considerations should be made when breaching in restricted terrain:

- OBSTINTEL collection is critical when conducting breaching operations in restricted terrain. It is essential that the breach force knows what types of obstacles are in place and knows the exact composition of any complex obstacle. With this information, the breach-force commander can ensure that the reduction assets necessary to reduce the obstacle are available and placed in the right order in the movement formation. The breach force's ability to move reduction assets within the formation may be extremely limited due to the terrain's constricting nature.
- Recon assets tasked to obtain OBSTINTEL should be prepared to recon different types of obstacles than would normally be found in open terrain. For example, in urban or mountainous terrain, the

enemy may use rubble obstacles, such as the *rock drop*. Additionally, recon assets are much more likely to encounter field-expedient or booby-trapped obstacles when conducting operations in restricted terrain. Recon assets used to collect OBSTINTEL must be able to recommend reduction methods to be used by the breach force when reducing the obstacle.

D-18. The proper application of the following breaching fundamentals is essential to the success of breaching operations conducted in restricted terrain. As with other breaching tenets, breaching fundamentals are normally applied differently in restricted terrain than in open terrain.

- **Suppress**. The need to prevent or degrade enemy fires and observation of friendly forces remains unchanged when breaching in restricted terrain. However, unlike a breaching operation conducted in open terrain in which the support force can suppress the enemy with long-range fires, massed indirect fires, and smoke, suppression will normally be achieved through close-in fighting. Forces tasked with suppressing the enemy may find it necessary to seize terrain at or on the far side of the obstacle to achieve their mission. In all probability, the fighting to achieve enemy suppression will be decentralized. Therefore, the ability to use indirect fires will be restricted due to the close proximity of friendly and enemy forces.
- **Obscure**. When trying to obscure the enemy with indirect fires, the same problems exist as when trying to suppress the enemy. Using mortars (because of their ability to fire high-level trajectory), smoke pots, and smoke grenades may be more effective. The potential for degrading friendly-force operations when using obscurants is greater because of the close-in nature of the fighting that occurs in restricted terrain.
- **Secure**. The responsibility to secure the reduction area may normally be given to the support force when conducting breaching operations in restricted terrain. Because the support force may find it necessary to seize terrain on the far side of the obstacle to suppress the enemy, securing the reduction area may be accomplished at the same time. When this situation occurs, the amount of maneuver forces within the breach force may be less than what is normally necessary when breaching in open terrain.
- **Reduce**. The ability to create more than one lane and to move reduction assets around each other may be severely restricted when breaching in restricted terrain. For this reason, accurate OBSTINTEL and a well-thought-out, well-rehearsed reduction plan are critical.
- **Assault**. As discussed earlier, the number of lanes available to the assault force may be less than what is actually needed. For example, a battalion-size assault force may be restricted to using a single lane through an obstacle due to the restricted terrain. For this reason, the assault-force commander must ensure that the sequencing of forces through the lane is appropriate to achieve the mission.
- **Mass**. Although mass remains a critical tenet for breaching operations in restricted terrain, the ability to achieve mass may be difficult and time-consuming. The assault force will attempt to

achieve a 3:1 combat-power ratio. However, due to the restricted nature of the terrain and the limited number of lanes that may be created, the assault force may take longer to get that much combat power to the far side of the obstacle. Units must consider ways to minimize the risk to the assault force as it builds its combat power on the far side of the obstacle.

- **Synchronize.** The reverse-planning process described in *Chapter 1* remains the key to successful breaching operations in restricted terrain. However, effective C^2, particularly the controlling of personnel and equipment around the point of breach, is critical for successful operations. The terrain's restricted nature magnifies mistakes made in bringing the proper reduction assets forward to the obstacle or in sequencing forces through created lanes. These mistakes result in increased time for the enemy to react to the breaching operation through repositioning and other means. For this reason, a well-rehearsed force is critical. All subordinate units must understand their mission, the overall timing of the operation, and how their actions relate to missions of other subordinate units.

CONSIDERATIONS FOR MILITARY OPERATIONS IN URBANIZED TERRAIN

D-19. Military operations in urbanized terrain (MOUT) is defined as all military actions that are planned and conducted on a topographical complex and its adjacent natural terrain where man-made construction is the dominant feature. Urbanized terrain affects the tactical options available to the commander and requires a thorough knowledge of unique terrain characteristics, detailed planning down to the smallest unit level, and sound leadership at all levels. The complexities of the urban environment, such as line-of-sight (LOS) restrictions, inherent fortifications, limited intelligence, densely constructed areas, and the presence of noncombatants, restricts current military technology. US forces do not possess the overwhelming technology advantages in an urban environment as in other environments.

D-20. A three-dimensional view of the terrain is essential to gain a complete understanding of observation and AAs. The close fight (inside the building) is difficult and decentralized, with control of the movement from room to room most often in the hands of squad and platoon leaders. Success in the movement from building to building is a combined-arms fight, which requires the active involvement of a battalion command post. Seeing the terrain in a three-dimensional view will identify opportunities for employment of all the battalion's combat power and enable the battalion to set the conditions for successful platoon and company assaults.

PLANNING

D-21. Close-quarters clearing techniques are designed to be executed by standard four-man breach and clearing teams. Therefore, the size of the breach organization will be significantly smaller than those involved when fighting occurs in open terrain. The reduction element must work with the breach leader to formulate the assault plan and to ensure that all planning considerations are completed.

D-22. The reduction element is responsible for the following tasks during the planning phase:

- Accessing all intelligence data on potential breach points.
- Helping the command element select the best method of entry (mechanical, shotgun ballistic breach, or explosive). (See *Figures D-7 through D-9*, and *Figure D-10*, page D-14)
- Assembling and preparing tools and equipment and constructing explosive entry charges and diversionary charges if required.
- Briefing team members.
- Rehearsing.
- Conducting a back brief of team members.
- Conducting a precombat inspection (PCI).

D-23. The commander, breacher, and other essential personnel assemble and analyze all available information, data, and intelligence pertaining to the intended target. This involves examining the design construction and material makeup of the structure to determine the best primary and alternate points of breach.

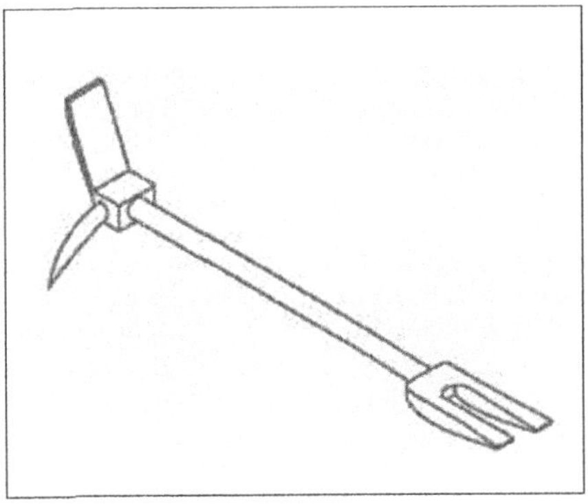

Figure D-7. Hooligan Tool

METHOD OF ENTRY

D-24. Once the point of breach has been identified, the reduction element leader continues to assemble and analyze all available information, data, and intelligence pertaining to the target structure. This involves a detailed examination of each point of breach's design, construction, and material makeup to determine the best primary, secondary, and tertiary breaching method. Again, tactics to be employed may dictate or rule out specific entry points.

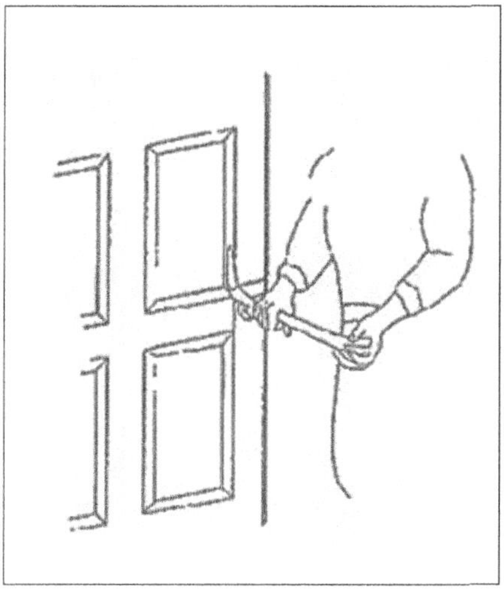

Figure D-8. Using a Hooligan Tool Against an Inward-Opening Door

Figure D-9. Attacking a Dead Bolt with a Sledgehammer

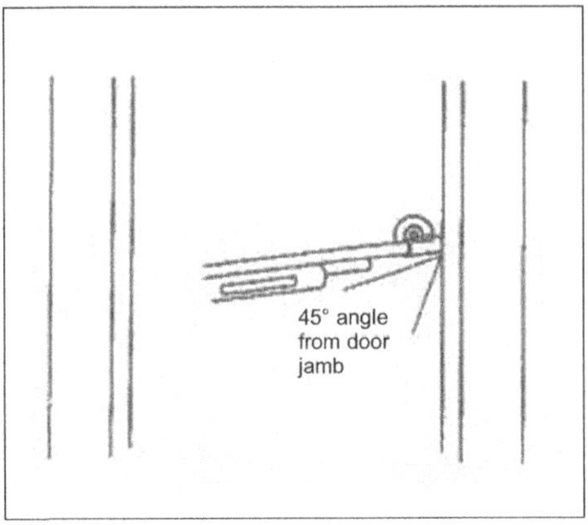

Figure D-10. Using a Shotgun Against a Door Lock

OBSTACLE INTELLIGENCE

D-25. Collecting the following OBSTINTEL is critical during MOUT:

- What type of mission is it?
- Are drawings or blueprints of the structure available?
- Are photographs of the point of breach and adjacent structures available?
- Can R&S personnel give a detailed description of the point of breach?
- Can counterintelligence personnel provide any information?
- Are there any human intelligence sources (agents, friendly forces, evacuees, prisoners, tourists, workers, businessmen, travelers) that can provide information relating to the point of breach?
- What type of shielding is available at potential breach points (natural or man-made)?
- How close will the assault force be positioned to the breach points?
- Do drawings of the target (if available) include gas, power lines, and other similar hazards?
- Is there a history of booby-trap use?
- Will the weather affect charge adhesion methods?

BREACHING FUNDAMENTALS

D-26. The proper application of the following breaching fundamentals is essential to ensure success in breaching conducted during MOUT. As with other breaching tenets, the breaching fundamentals are normally applied differently in urbanized terrain than in open terrain.

- **Suppression**. Maneuvering to achieve suppression of the enemy will be decentralized. Therefore, the ability to use indirect fires will be restricted due to the close proximity of friendly and enemy forces and civilians.

- **Obscure**. The potential for degrading friendly-force operations with obscurants is greater due to the close-in fighting that occurs in built-up terrain. Hand-emplaced smoke will likely be the most effective.
- **Secure**. Breach and clearing teams secure a foothold near the point of breach. Local security must consider all possible locations for the enemy to hide.
- **Reduce**. Redundant systems are critical during MOUT. The reduction element is responsible for preparing equipment/charges for emplacement, searching for improvised explosive devices (IEDs), and verifying target intelligence.
- **Assault**. The actual attack of the objective using close-quarters battle techniques should be conducted according to the assault plan. The unit should be prepared to conduct additional planned and unplanned breaches.

Appendix E

Route Clearance

Route clearance is a combined-arms operation typically executed by an infantry company or a cavalry troop. Units must clear lines of communication (LOC) of obstacles and enemy activity that disrupt battle-space circulation. Units must conduct route clearance to ensure that LOC enable safe passage of combat, combat-support (CS), and CSS organizations. Clearance operations are normally conducted in a low-threat environment.

METHODS AND LEVELS

E-1. The information gathered from the IPB and the recon effort determines the method and type of route clearance to conduct. The determination is based on the situation, the time available, the threat level, and available assets.

E-2. During stability and/or support actions, it is recommended that former-warring-faction (FWF), engineer-equivalent clearance teams precede US clearance teams within the FWF's AO. Do not assume that FWF clearance teams will be thorough in their clearance operation. Treat the route as unsafe until US or allied force clearance teams have proofed the route to confirm that it is cleared.

CLEARANCE METHODS

E-3. There are three methods of route clearance—linear, combat, and combat route. The method employed depends on the situation, the time available, and the clearance assets available. The maneuver force should always establish static security positions at critical locations following the completion of route clearance.

- **Linear**. In linear clearance (*Figure E-1,* page E-2), the clearance begins at checkpoint 1 and is completed at checkpoint 2. This method provides the best assurance of route clearance. Although this is an effective method, it is not the most secure method in a high-threat environment. It is also time-intensive and constrains the maneuver commander's flexibility.
- **Combat**. Whereas linear clearance focuses on a specific route, combat clearance (*Figure E-2,* page E-2) focuses on specific points along a route. As mentioned previously, the IPB can identify likely areas for obstacle and ambush locations. These areas become NAIs or objectives for combat clearance missions. The combat clearance method divides a route into sections according to the number of suspected high-threat areas. Once the unit clears these areas, the route is considered clear. Combat forces can patrol the route from these objectives to ensure

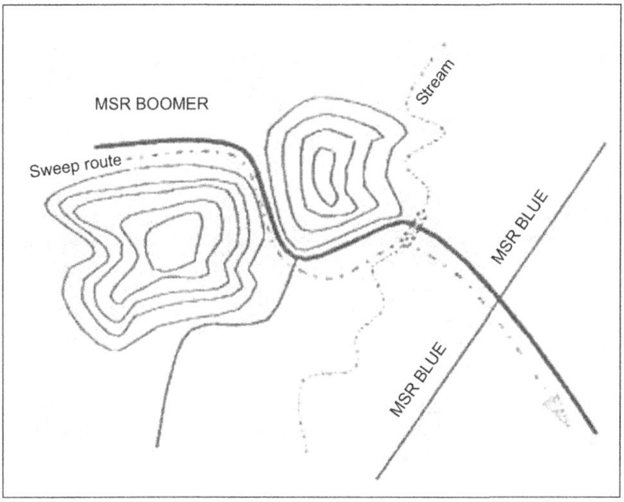

Figure E-1. Linear Clearance Method

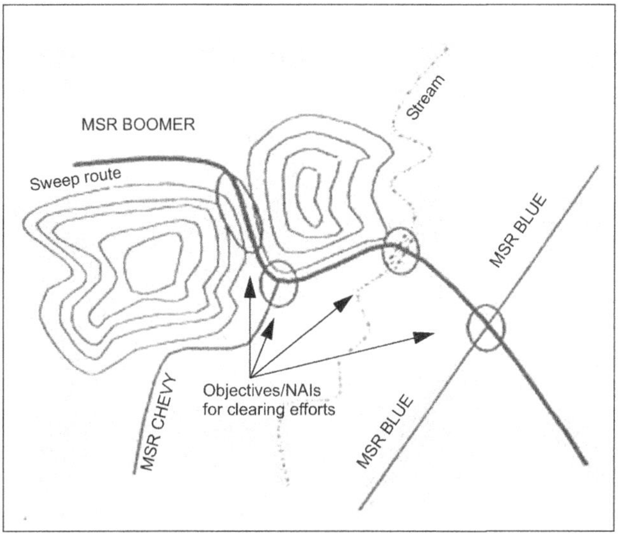

Figure E-2. Combat Clearance Method

that the route is secure and, if necessary, the sweep element can sweep the surrounding area if a minefield is found. Following the seizure of these objectives, the commander must assume a moderate risk that the S2 and the force engineer have identified all high-threat areas and that the route is clear of obstacles.

- **Combat route**. This method combines the clearance capabilities of the linear method with the security and surprise elements of the combat method. It is a two-phase, force-intensive operation and may require a battalion-size element, depending on the route's length. First, identified high-threat areas are targeted, secured, and cleared of obstacles and enemy forces before a sweep team moves. The sweep team then travels down the road and clears obstacles missed or not identified during the IPB/engineer battlefield assessment (EBA). The advantage of this method is that the TF commander immediately secures MSRs and then finds the enemy, confident that MSRs are relatively safe.

SWEEP LEVELS

E-4. The four levels of sweep operations are discussed below. All four levels can be modified to meet the unit's time and equipment limitations, but the commander assumes greater risk when the clearance level is modified. All four sweep levels can be used with any of the three clearance methods (linear, combat, and combat route).

- **Level one**. This is the fastest, riskiest type of route sweep. It is ideal for an armored/mechanized team. The sweep relies primarily on visual detection, using thermal or infrared sights or the naked eye for mine identification. It is followed immediately by a secondary, mechanical detection system such as MCRs. The sweep team is a squad-size or larger element that is task-organized with mine detectors, demolitions, and possibly an interim vehicle-mounted mine detector (IVMMD). The sweep team looks for minefields along the road width of a route. Security and support teams consist of a maneuver platoon to provide overwatching fire. The primary objective of a level one sweep is speed, with the sweep team moving at 5 to 8 miles per hour. The sweep team identifies immediate risks to traffic, neutralizes those risks, and continues on with the mission. If mine rollers are not available, a sandbagged 2 1/2-ton truck driven backwards can be used for a level one sweep, but only as a last resort. In this situation, the mine rollers are only a means of detection, not breaching. Mine rakes or plows are not always satisfactory substitutes because they most likely will destroy road surfaces.
- **Level two**. A level two sweep uses electronic measures as the primary detection method in high-threat areas. These areas include intersections, choke points, and areas within 10 meters of wood lines. A level two sweep employs more caution and forces the unit to update the IPB before beginning the mission. This level of operation employs a company team for security and C^2. Dismounted forces clear and secure the flanks and the far side of an identified minefield, while an engineer squad clears the road area.

- **Level three**. A level three sweep is more in-depth and more time-consuming. The sweep team must examine the route's entire width, including the shoulders and ditches. This technique ensures that follow-on forces are protected if they are forced to the side of the road. The security and support element also moves, mounted or dismounted, to provide rapid response and security. The sweep team provides a route recon report that updates current maps and further identifies high-threat areas. Staffs gain valuable information from this report for future operations.
- **Level four**. A level four sweep is very time-consuming. It relies on visual and electronic means for clearance. The platoon-size sweep team is dismounted to focus attention on the route's entire length. The company-size security element clears and secures the flanks and front of the sweep team (at least 100 meters in each direction in forested areas). This clears the area of off-route and command-detonated mines and allows the team to focus solely on the route. MCRs are used to proof the route after the sweep team passes through the area. Level four sweeps include route and area recons, and they are used when thoroughness and security, not speed, are critical to the mission.

EQUIPMENT

E-5. The IVMMD and the MinFlail are used to conduct route clearance.

INTERIM VEHICLE-MOUNTED MINE DETECTOR

E-6. The IVMMD (*Figure E-3*) is used in all levels of conflicts and stability and/or support actions. It is mounted on a blast- and fragmentation-survivable vehicle and is designed to detect and mark buried and surface-laid, metallic AT mines. The primary mission of the IVMMD is to detect mines during route-clearance operations. The system should not be operated in an environment where the enemy employs mines that are not pressure-fused.

System Components

E-7. A complete IVMMD consists of one mine-detection vehicle (MDV), one towing/mine-detection vehicle (T/MDV), three mine-detonation trailers, a spare-wheel module for the MDV, a spare-wheel module for the T/MDV, and a container of spare parts.

E-8. The MDV's only mission is to detect mines. It can negotiate vertical slopes up to a 20 percent grade. The MDV employs a 4-cylinder engine and a manual transmission to propel the 4.8-ton vehicle with a 3-meter-wide detection array. The detection array consists of two separate induction coils (one for the left side and one for the right side) that detect magnetic fields below the vehicle. The detection array is suspended between the two axles of the vehicle. When the detector encounters a metallic object, the operator is notified by an audible signal in the earphone and a visual signal on the dashboard that denotes which side array detected the object. The operator then stops the vehicle, backs it up, and reencounters the metallic object. (The MDV has two detection modes—the locate mode is used to identify the metal object, and the pinpoint mode is used to find the center of the object.) When

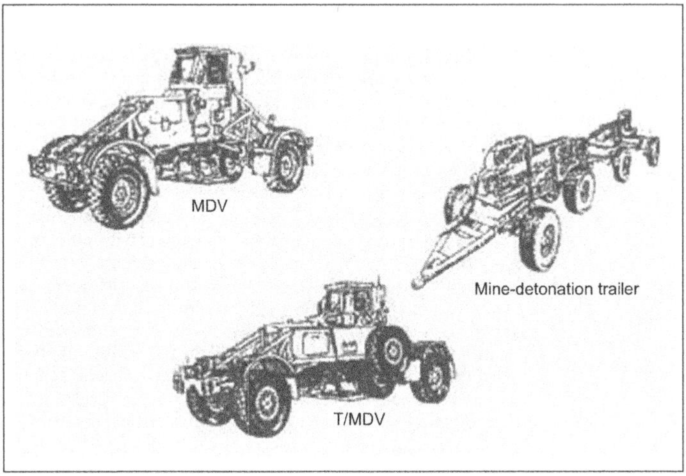

Figure E-3. IVMMD Components

the operator encounters the strongest signal, he activates the marking system (a nozzle mounted on the rear frame and centered on each detection array) that deploys a water-based ink onto the roadway.

E-9. The MDV is designed to produce very little ground pressure; however, it may detonate pressure-fused mines, depending on their fuse sensitivity. The MDV produces 27.9 pounds per square inch (psi) of ground pressure when the tires are inflated to 14.5 psi and 21.8 psi of ground pressure when the tires are inflated to 8.7 psi.

E-10. The T/MDV has a 6-cylinder engine and the same detection and marking system as the MDV. The T/MDV produces 49.8 psi of ground pressure when the tires are inflated to 29 psi and 23.4 psi of ground pressure when the tires are inflated to 8.7 psi. The T/MDV tows three mine-detonation trailers, and the recommended maximum operating speed while towing the trailers is 20 kph. The T/MDV (with trailers) can negotiate vertical slopes up to a 20 percent grade; however, going down such slopes is difficult. The T/MDV must be in first gear, and the trailer brakes must be deployed to decrease the speed of the vehicle when going down a slope.

E-11. The mine-detonation trailers are very heavy and specifically designed to apply heavy ground pressure that initiates pressure-activated mines. Each trailer has two axles of different lengths so that the three trailers provide a full 3-meter-wide proofing capability behind the T/MDV. If a mine detonates underneath the trailers, the wheel bolts are designed to sheer so that repair is limited to replacing a single wheel.

Employment Concept

E-12. The IVMMD is used to support route-clearance operations. Clearance operations ensure that LOC are safe for the passage of personnel and equipment. The IVMMD should not be used during hours of limited visibility, because it hampers the operator's ability to see surface-laid mines and visual signatures that indicate mining activities.

MINIFLAIL

E-13. The MiniFlail (*Figure E-4*) is a remotely operated, LOS, AP-mine and unexploded-ordnance (UXO) neutralization system that was developed for use by US light forces. The MiniFlail detonates or disables AP mines from a safe operating distance. It neutralizes by striking objects with a rotating chain assembly, called a flail, and clears a footpath about 1.1 meters wide. The system neutralizes AP mines and UXO by detonation, mechanical destruction, or displacement from the cleared lane. The MiniFlail is about 1.3 meters wide, 1.3 meters tall, and 3 meters long; it weighs 1,100 kilograms. The system is operated by a hand-held controller that has a maximum range of 300 meters. It is fully armored with a material similar to Kevlar, and the tires are filled with foam. The MiniFlail is a self-articulating, hydraulically powered shaft with 84 chains; each chain is 0.5 meter long.

Figure E-4. MiniFlail

PLANNING

E-14. The principles of breaching operations apply to developing and executing the route-clearance mission. The breaching tenets (intelligence, fundamentals, organization, mass, and synchronization) are the basis for planning.

INTELLIGENCE

E-15. Incorporating the IPB and METT-TC factors into route-clearance operations enables units to predict what the enemy will do and where it will do it. After the S2 and the engineer identify the most probable threat sites, the S2 designates them as NAIs. These NAIs are the focus of the recon effort.

Recon assets confirm the presence or absence of ambushes, UXO, and obstacles. The information gathered from the IPB and the recon effort determines the method and type of route clearance necessary. It also helps the commander determine any outside resources (explosive ordnance disposal [EOD] and special-operations forces [SOF]) that he may need.

FUNDAMENTALS

E-16. SOSRA may not be executed, but it is planned as it is in breaching operations. Units must be prepared to execute SOSRA fundamentals as necessary.

ORGANIZATION

E-17. Route-clearance operations are similar to breaching operations; therefore, commanders should plan and coordinate breaching fundamentals to suppress, obscure, secure, and reduce obstacles. Task-organizing for route clearance is also similar to that required for breaching operations. The assault force becomes the security element, the breach force becomes the sweep element, and the support force remains the same.

E-18. The significant difference between breaching and clearing operations is that breaching usually occurs during an attack (while under enemy fire) to project combat power to the far side of an obstacle. Route clearance focuses on opening LOC to ensure the safe passage of combat and support organizations within an AO. *Table E-1*, page E-8, shows a sample task organization for a route clearance.

MASS

E-19. Sufficient maneuver and engineer assets must be allocated to the sweep force. The length and width of the route and the type of clearance to be conducted determine the composition of the breach force. Clearing a Class A military road with the deliberate sweep technique requires at least two IVMMDs and/or two engineer squads due to the total lane width to be cleared and the requirement for rotating mine-detector operators. Depending on the type of sweep operation, the commander should plan for a 50 percent loss of sweep assets.

SYNCHRONIZATION

E-20. All aspects of synchronization should be implemented when planning route clearance. It is especially important that the clearance force conduct rehearsals at the combined-arms level. Rehearsals should include—

- Reaction to enemy contact, including ambushes.
- Communications exercise.
- Fire support (obscuration smoke, immediate suppression fires, CFZs, and no-fire area around the clearance site).
- CSS (casualty evacuation, marking materials, and demolitions resupply).

Table E-1. Sample Task Organization for a Route Clearance

Team	Support Force	Security Force	Sweep Force
Heavy	Mechanized infantry platoon with dismount capability Armor platoon FIST/COLT/Striker ADA section	Mechanized infantry platoon Engineer squad Mortar section Medical team (two ambulances) PSYOP team FIST MP element	Engineer platoon with organic vehicles Armor platoon with plows and rollers
Light/Heavy	Two infantry platoons (light) FIST/COLT/Striker ADA section	Bradley platoon with dismount capability Engineer squad 60-mm mortar section Medical team (two ambulances) PSYOP team FIST MP element	Engineer platoon with organic vehicles Armor platoon with plows and rollers
Light	Two infantry platoons (light) FO/COLT/Striker ADA section	AT/MP section with M60/MK19 mix 60-mm mortar section Medical team (two ambulances) PSYOP team FO MP element	Engineer squad (+) Infantry platoon (light) AT/MP section with M60/ MK19 mix

PLANNING CONSIDERATIONS

E-21. The purpose of breaching is to project combat power to the far side of an obstacle, and breaching usually occurs under enemy fire. Route-clearance operations focus on opening and maintaining LOC to ensure the safe passage of combat, CS, and CSS organizations. Like breaching, route-clearance operations require extensive Battlefield Operating System (BOS) coordination. The planning considerations shown in *Table E-2* should be used by brigade and battalion TFs when planning route-clearance operations. The obstacle template produced from this process determines which method and type of route clearance is required.

Table E-2. Planning Considerations for Route Clearance

INTELLIGENCE
Identify choke points, bridges, tunnels, critical road junctions, and built-up areas. These are the most likely locations for obstacle emplacement. However, depending on the enemy's overall mission, it may not always emplace obstacles at these locations. This is especially true if the enemy's goal is to psychologically disrupt convoys.
Maintain a situation map that reflects the most current intelligence information.
Maintain an incident map with a graphics overlay to facilitate a pattern analysis.
Maintain a threat order-of-battle database, such as how the enemy has disrupted LOC in the past.
Develop a detailed ISR plan that incorporates modern battle-space techniques and systems, such as ground sensors, forward-looking airborne radar, and satellite images. As a minimum— • Coordinate for UAV support, if available. • Develop infiltration routes to support recon and security at likely enemy ambush sites. • Develop an estimate of the impact to COBs (local nationals, NGOs, and PVOs). • Conduct a daily flight over the area to provide up-to-the-minute intelligence. • Coordinate with the USAF to periodically check the route (for example, use an AC-130 Specter gunship).
Establish liaison with the host nation, NGOs, and SOF.
Provide detailed OBSTINTEL, including— • Description of mines or explosive devices most likely encountered. • Composition and pattern of obstacles. • Enemy actions or techniques used during obstacle emplacement.
Provide intelligence updates on mine-hazard areas to company team leaders before departure. This should be in the form of a 1:50,000 enemy SITEMP overlay (confirmed and suspected/templated).
MANEUVER
Provide personnel for TCPs.
Secure flanks (at least 100 meters) and the far side of suspected and known obstacle locations.
Close the route to US-controlled traffic during route-clearance operations to minimize the target presented to enemy forces.
Identify and clear potential sniper positions before beginning obstacle reduction or clearance.
Provide security for the cleared route.
Plan the building of static security points along the cleared route to reduce the probability of reseeding.
FIRE SUPPORT
Plan smoke for templated obstacle locations.
Position mortars to ensure continuous coverage of the operation (one technique is to move under the control of the support force).
Prepare fires within the tactical rules of engagement.
Designate obstacle clearance sites as CFZs and no-fire areas.
Plan suppression fires on enemy elements capable of placing direct or indirect fires on the points of breach.

Table E-2. Planning Considerations for Route Clearance (continued)

MOBILITY/SURVIVABILITY
Report, clear, and mark obstacles and explosive devices to facilitate unimpeded movement.
Ensure that lane marking meets the standards directed by the supported commander and that materials and patterns are standard throughout the route.
Consider including road-repair equipment and material as part of the breach force (for example, a 5-ton dump truck filled with soil and an ACE to spread the soil).
Keep all radios, electronic equipment, and aviation assets at a safe distance during reduction operations.
Block uncleared roads and trails that branch from the route being cleared. This protects units from inadvertently traveling an uncleared route.
Debrief the chain of command and the TF S2 on the location, composition, and orientation of all obstacles cleared and encountered. This assists the S2 in IPB pattern analysis.
Provide detailed OBSTINTEL on minefields, including— • Description of mines or explosive devices most likely to be encountered. • Composition and pattern of obstacles. • Enemy actions or techniques used during obstacle emplacement.
ADA
Assist the S2 with the threat-air portion of the IPB.
Use the following passive air-defense measures: • Eliminate glare by using mud, tape, cardboard, or camouflage nets to cover headlights, mirrors, and portions of windshields. • Reduce dust clouds by reducing speed. • Plan routes that offer natural concealment. • Use air guards.
Increase the distance between vehicles.
Incorporate Stinger missile teams into the support force.
CSS
Ensure that clearance operations are supported by a logistical/CHS package.
Plan for air and ground evacuation of casualties. The preferred evacuation method is by air; the routine method is by ground.
Conduct an air-mission brief with air-ambulance assets, including pickup zones and markers.
Rehearse procedures for evacuation requests.
Ensure that the medical team consists of one or two ambulances.
Locate the medical team with the support force.
Identify the ambulance exchange point along the route to be cleared.
Ensure that appropriate personnel wear flak vests or IBASIC (*Figure E-5*, page E-12).
Ensure that all vehicles have tow cables in the front and the rear, and plan recovery assets for extraction purposes.
Ensure that all vehicles carrying troops have hardening (sandbags on floors and sides).
Provide MP and explosive-sniffing dogs to help in clearance operations and provide security for convoys during and after clearing operations.

Table E-2. Planning Considerations for Route Clearance (continued)

C^2
Designate a controlling, coordinating, and supporting HQ for route movement.
Designate, recognize, and include the following minefield indicators as part of company-team rehearsals: • Dead animals. • Avoidance by the local population. • Signs of digging. • Signs of concrete removal. • Disturbances in a road, such as holes or grooves. • Boxes or parcels placed along the road/shoulder. • Parked vehicles or bicycles without operators. • Wire on the road surface or extending onto the shoulders. • Evidence of vegetation disturbance along shoulders. • Evidence of mine-peculiar supplies such as wrenches, shipping plugs, wrapping paper or safety collars from fuses. • Posted signs that covertly alert the local population to the presence of mines. • Disturbances in tire tracks.
Designate a reserve force (at least platoon-size) that is mechanized or air-assault capable.
Ensure that proper rehearsals are planned and conducted. As a minimum, the unit should rehearse actions at the obstacle, actions on enemy contact, casualty evacuation, and control of COBs.
Ensure that the unit has a clear understanding of the mission, the intent, and the end state. For example, the clearing-unit commander should understand that his unit must clear the road width, including the shoulders, and secure the route.
Ensure that the unit— • Controls the movement of all personnel and equipment along the route. • Clears personnel within the minimum safe distance when clearing mines. • Prepares a risk assessment of the mission before issuing the OPORD. • Tracks the status of routes based on the amount of time since the route was cleared and the intelligence and enemy situations. • Tracks the progress of the clearance operation and integrates it into the maneuver and CSS plans. • Determines the route length, using clearly definable start and end points. • Coordinates with adjacent units, the host nation, NGOs, PVOs, and SOF. • Sets priorities for the route-clearance element.
SPECIAL OPERATIONS
Ensure that PSYOP/CA support the counterintelligence effort by conducting civilian interviews.
Direct civilians along the MSR to the displaced-personnel holding areas or along the routes that the unit has indicated for use.
Employ PSYOP/CA teams forward to disperse civilians and provide traffic management to isolate the route during clearance operations.

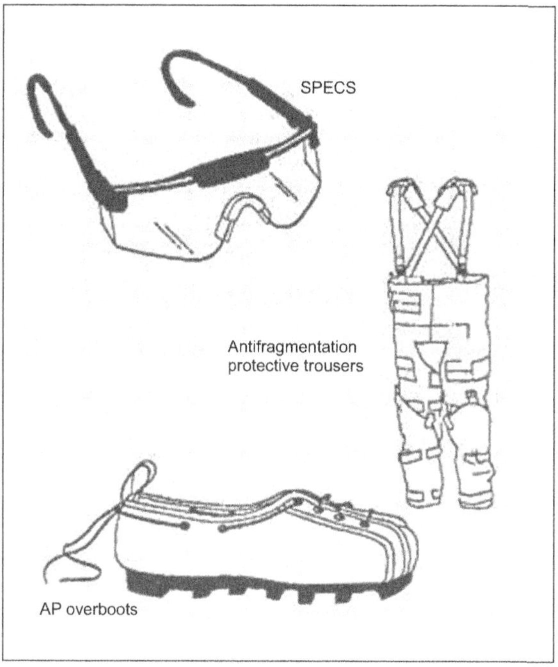

Figure E-5. IBASIC

TASK ORGANIZATION

E-22. A brigade or battalion TF normally conducts clearance-in-zone operations. To clear a route, the battalion TF focuses a company team as the main effort on the proposed MSR. *Table E-1*, page E-8, shows a sample task organization for a route clearance.

SUPPORT FORCE

E-23. The support force provides security for the breach force and neutralizes hostile forces. In rugged terrain or highly mined areas, moving the support force on the flanks may be too risky. In this situation, aviation assets may be used to provide flank security while ground forces provide rear security. The support force may also be tasked to search for suspected off-route mines.

SECURITY FORCE

E-24. The security force is prepared to maneuver and destroy enemy forces that the support force cannot neutralize. The security force may be comprised of maneuver and engineer forces augmented with a medical team, a psychological operations (PSYOP) team, and an EOD team. Additionally, the security force may be supported by a mortar section.

SWEEP FORCE

E-25. The sweep force is comprised of maneuver and engineer forces with detection equipment. The sweep force sweeps the route and reduces mine and explosive threats. As in breaching operations, the sweep force consists of a security element and a reduction element. The security element provides local security for the sweep force. The reduction element consists of sweep teams (personnel and equipment) to reduce obstacles and clear mines.

E-26. A sweep team is a trained detection team that searches for mines and explosive devices. The organization of the sweep team depends on the type of sweep mission, the length and width of the road to be cleared, the time and equipment available, and the surface composition (pavement, gravel, dirt) of the road to be swept. A dismounted platoon-size element can normally clear a 4.5-meter-wide path, and a squad-size element can normally clear a 1.5-meter-wide path. Mounted detection equipment, such as the IVMMD, can detect mines along a 3-meter-wide path. See *FM 20-32* for detailed information on the composition of a sweep team.

Glossary

AA avenue of approach

ACE armored combat earthmover

AD armored division

ADA air-defense artillery

ADAM area-denial artillery munition

AHD antihandling device

ammo ammunition

AO area of operations

AP antipersonnel

APC armored personnel carrier

APOBS Antipersonnel Obstacle Breaching System

ASLT assault

AT antitank

atk attack

ATTN attention

AVLB armored vehicle-launched bridge

AVLM armored vehicle-launched MICLIC

BFV Bradley fighting vehicle

BHL battle handover line

BOS battlefield operating system

C^2 command and control

CA civil affairs

CAS close air support

CCIR commander's critical-information requirements

cdr commander

CFZ critical friendly zone

CHS combat health support

CLAMS Cleared-Lane Marking System

cm centimeter(s)

cmd command

COA course of action

COB civilians in the battle space

COLT combat observation lasing team

CP checkpoint

CS combat support

CSS combat service support

Czech Czechoslovakia

DA Department of the Army

det detonated

dia diameter

DPICM dual-purpose improved conventional munition

DS direct support

DST decision-support template

ea each

EA engagement area

EBA engineer battlefield assessment

EOD explosive ordnance disposal

EOM end of mission

FD firing device

FECS Field-Expedient Countermine System

FFIR friendly forces information requirements

FIST fire-support team

FM field manual

FM frequency-modulated

FO foward observer

FRAGO fragmentary order

FSE fire-support element

FSO fire-support officer

FWF former warring faction

GPS Global Positioning System

HE high explosive

HEMMS Hand-Emplaced Minefield Marking System

HEMTT heavy expanded mobility tactical truck

how howitzer

HQ headquarters

IBASIC Improved Body Armor System, Individual Countermine

ICP intelligence-collection plan

ID infantry division

IDA improved dog-bone assembly

IED improvised explosive device

IFV infantry fighting vehicle

IP improved plow

IPB intelligence preparation of the battlefield

IR information requirements

ISR intelligence, surveillance, and reconnaissance

IVMMD interim vehicle-mounted mine detector

km kilometer(s)

kph kilometers per hour

lb pound(s)

LD line of departure

ldr leader

LO liaison officer

LOC lines of communication

LOS line of sight

m meter(s)

MCAP mine-clearing/armor-protection

MCB mine-clearing blade

MCR mine-clearing roller

MCRP Marine Corps reference publication

MDMP military decision-making process

MDV mine detection vehicle

METT-TC mission, enemy, terrain, troops, time available, and civilian considerations

MIBN motorized infantry battalion

MIC motorized infantry company

MICLIC mine-clearing line charge

MIP motorized infantry platoon

MLC military load class

mm millimeter(s)

MOUT military operations in urbanized terrain

MP military police

MSR main supply route

NA not applicable

NAI	named area of interest
NATO	North Atlantic Treaty Organization
NBC	nuclear, biological, chemical
NGO	nongovernment organization
NSN	national stock number
OBJ	objective
OBSTINTEL	obstacle intelligence
OP	observation post
OPFOR	opposing force
OPORD	operation order
oz	ounce(s)
PCI	precombat inspection
PIR	priority intelligence requirements
PL	phase line
psi	pounds per square inch
PSN	position
PSYOP	psychological operations
PVO	private volunteer organization
R&S	reconnaissance and surveillance
RAAM	remote antiarmor mine
RCU	remote control unit

RL release line

RP reference point

S2 Intelligence Officer (US Army)

S3 Operations and Training Officer (US Army)

SBF support by fire

SCATMINE scatterable mine

SITEMP situation template

SOF special-operations forces

SOP standing operating procedures

SOSRA suppress, obscure, secure, reduce, and assault

SP start point

SPECS Special Protective Eyewear Cylindrical System

SPOTREP spot report

STANAG Standardization Agreement

T/MDV towing/mine-detection vehicle

TAI targeted area of interest

TC tank commander

TCP traffic-control post

TF task force

TM technical manual

TOC tactical operations center

TRADOC United States Army Training and Doctrine Command

TRP target reference point

TTP tactics, techniques, and procedures

UAV unmanned aerial vehicle

UK United Kingdom

UMCP unit maintenance collection point

US United States

USAF United States Air Force

UXO unexploded ordnance

WLGH weapon-launched grapnel hook

WO warning order

wt weight

Yug Yugoslavia

Bibliography

DA Form 2028. *Recommended Changes to Publications and Blank Forms.* 1 February 1974.

FM 3-50. *Smoke Operations.* 4 December 1990.

FM 5-71-2. *Armored Task-Force Engineer Combat Operations.* 28 June 1996.

FM 5-71-3. *Brigade Engineer Combat Operations (Armored).* 3 October 1995.

FM 5-71-100. *Division Engineer Combat Operations.* 22 April 1993.

FM 5-105. *Topographic Operations.* To be published in the next six months.

FM 5-170. *Engineer Reconnaissance.* 5 May 1998.

FM 6-71. *Tactics, Techniques, and Procedures for Fire Support for the Combined Arms Commander.* 29 September 1994.

FM 6-121. *Tactics, Techniques, and Procedures for Field Artillery Target Acquisition.* 25 September 1990.

FM 7-10. *The Infantry Rifle Company.* 14 December 1990.

FM 17-98. *Scout Platoon.* 9 September 1994.

FM 20-32. *Mine/Countermine Operations.* 29 May 1998.

FM 34-130. *Intelligence Preparation of the Battlefield.* 8 July 1994.

FM 71-1. *Tank and Mechanized Infantry Company Team.* 26 January 1998.

FM 71-2. *The Tank and Mechanized Infantry Battalion Task Force.* 27 September 1988.

FM 71-3. *The Armored and Mechanized Infantry Brigade.* 8 January 1996.

FM 90-7. *Combined Arms Obstacle Integration.* 29 September 1994.

FM 100-5. *Operations.* 14 June 1993.

FM 100-7. *Decisive Force: The Army in Theater Operations.* 31 May 1995.

FM 101-5. *Staff Organization and Operations.* 31 May 1997.

FM 101-5-1. *Operational Terms and Graphics (MCRP 5-2A).* 30 September 1997.

STANAG 2036. *Land Mine Laying, Marking, Recording and Reporting Procedures.* Edition 5. 28 January 1999.

TM 5-6665-298-10. *Operator's Manual for AN/PSS-12 Mine Detecting Set (TM 5-6665-298-10).* 28 February 1995.

TM 9-1375-215-14&P. *Operator's, Unit, Direct Support and General Support Maintenance Manual (Including Repair Parts and Special Tools List) for Demolition Kit, Mine Clearing Line Charge (MICLIC)*. 31 January 1992.

Index

Made in the USA
Las Vegas, NV
30 August 2026

53973056R00144